PLANNING AND DESIGN
OF BRIDGES

PLANNING AND DESIGN OF BRIDGES

M. S. TROITSKY, D.Sc.
Professor Emeritus of Civil Engineering
Concordia University
Montreal, Canada

JOHN WILEY & SONS, INC.
New York / Chichester / Brisbane / Toronto / Singapore

This text is printed on acid-free paper.

Copyright © 1994 by John Wiley & Sons, Inc.

All rights reserved. Published simultaneously in Canada.

Reproduction or translation of any part of this work beyond that permitted by Section 107 or 108 of the 1976 United States Copyright Act without the permission of the copyright owner is unlawful. Requests for permission or further information should be addressed to the Permissions Department, John Wiley & Sons, Inc., 605 Third Avenue, New York, NY 10158-0012.

This publication is designed to provide accurate and authoritative information in regard to the subject matter covered. It is sold with the understanding that the publisher is not engaged in rendering legal, accounting, or other professional services. If legal advice or other expert assistance is required, the services of a competent professional person should be sought.

Library of Congress Cataloging in Publication Data:
Troitsky, M. S.
   Planning and design of bridges/M. S. Troitsky.
      p.   cm.
   Includes bibliographical references and index.
   ISBN 0-471-02853-3
   1. Bridges—Design and construction. 2. Bridges—Planning.
I. Title.
TG300.T76   1994
624'.25—dc20   94-6946

Printed in the United States of America

10 9 8 7 6 5 4 3 2

To
My sisters Anna and Helen
and
My brother Victor

CONTENTS

PREFACE xvii

1. HISTORY OF BRIDGES 1

    1.1   Introduction / 1
    1.2   Development of Bridge Types / 2
    1.3   The Ancient Period / 5
    1.4   The Roman Period / 9
    1.5   The Middle Ages / 13
    1.6   An Age of Iron and Steel Bridges / 17
    1.7   An Era of Suspension Bridges / 22
    1.8   An Era of Cable-Stayed Bridges / 30
    1.9   An Era of Concrete Bridges / 32
           1.9.1 Reinforced Concrete Bridges / 32
           1.9.2 Prestressed Concrete Bridges / 34
    1.10 Proposed Future Bridges / 36
           1.10.1 Messina Straits Suspension Bridge / 36
           1.10.2 Normandy Cable-Stayed Bridge / 37
           1.10.3 Akaghi Suspension Bridge, Japan / 38
           1.10.4 Bering Strait Bridge Project / 39
           1.10.5 Proposed Gibraltar Strait Bridge / 41
    References / 42

2. BRIDGE LOCATION 45

2.1 Introduction / 45
2.2 Bridges in Urban Regions / 45
    2.2.1 Transportation Requirements / 45
    2.2.2 Technical Requirements / 46
    2.2.3 Aesthetic Considerations / 46
2.3 The Optimum Location of a Bridge Considering Traffic Flow / 46
2.4 Numerical Example / 50
References / 52

3. BRIDGE LAYOUT 53

3.1 Square or Skew Bridge / 53
3.2 Angle of Crossing / 54
    3.2.1 Skewed Substructure / 56
    3.2.2 Alternative Layouts of the Bridge / 57
    3.2.3 Summary / 58
3.3 Bridge Approaches / 59
3.4 Site Investigation / 59
    3.4.1 General data / 59
    3.4.2 Office work / 60
    3.4.3 Field work / 60
References / 61

4. CROSSING THE RIVER 62

4.1 Laying out the Bridges Spans / 62
4.2 Principles of Bridge Crossing Design and its Structures / 63
    4.2.1 General Data on Water Passage and Bridge Crossing / 63
    4.2.2 Exploration of Bridge Crossing / 65
    4.2.3 Choice of Location for Bridge Crossing and Design of Bridge Opening / 66
    4.2.4 Sequence of Bridge Design and Comparison of Alternatives / 67
    4.2.5 Bridge Approaches—Regulation and Bank Protection Structures / 68
4.3 General Conditions for the Layout of a Bridge Across a River / 69

4.4 Computing Openings of Bridges / 73
    4.4.1 Definitions / 73
    4.4.2 Computation of Bridge Opening / 74
4.5 Scour / 76
    4.5.1 Local Scour / 76
    4.5.2 Factors Affecting Scour / 77
    4.5.3 Protection of Foundation Against Scour / 79
    4.5.4 Suggestions to Minimize Scour Effects / 80
    4.5.5 Estimate of Possible Scour / 81
    4.5.6 Computation of the Bridge Opening when Scour or Erosion is not Permitted / 81
    References / 84

5. STRUCTURAL BRIDGE SYSTEM   86

5.1 Superstructure and Substructure / 86
5.2 Bridge Geometry / 88
5.3 Determining the Most Economical Bridge Span / 92
5.4 Bridge Types / 95
    References / 96

6. SUPERSTRUCTURE—STEEL BRIDGES   97

6.1 Rolled-Beam Bridges / 97
6.2 Plate-Girder Bridges / 98
6.3 Composite Beam and Girder Bridges / 99
6.4 Continuous Composite-Plate Girder Bridges / 100
    6.4.1 Introduction / 100
6.5 Optimum Height of the Plate Girder / 101
6.6 Helpful Hints for Girder Bridge Design / 104
6.7 Composite Box-Girder Bridges / 105
6.8 Orthotropic Deck Bridges / 106
    6.8.1 Box Girders / 107
    6.8.2 Cable-Supported Box Girders / 107
    6.8.3 Ribs / 107
    6.8.4 Floor Beams / 108
    6.8.5 Orthotropic Deck Surfacing / 109
6.9 Truss Bridges / 109
6.10 Optimum Height of the Truss / 111

6.11 Arch Bridges / 112
    6.11.1 General Characteristics of the Arches / 112
    6.11.2 Basic Types of Arches / 113
    6.11.3 Abutments / 114
    6.11.4 Structural Characteristics / 114
    6.11.5 Deck Construction / 114
6.12 Suspension Bridges / 115
    6.12.1 Stiffening Girders or Trusses / 115
    6.12.2 Cable Systems / 115
    6.12.3 Towers / 116
    6.12.4 Floor System / 116
    6.12.5 Continuity / 117
    6.12.6 Suspension Systems / 117
6.13 Cable-Stayed Bridges / 118
    6.13.1 Arrangement of the Stay Cables / 118
    6.13.2 Positions of the Cables in Space / 119
    6.13.3 Tower Types / 120
    6.13.4 Deck Types / 121
    6.13.5 Main Girders and Trusses / 121
    6.13.6 Structural Advantages / 124
    6.13.7 Comparison of Cable-Stayed and Suspension Bridges / 125
    6.13.8 Composite Cable-Stayed Bridges / 126
    References / 126

7. SUPERSTRUCTURE–REINFORCED CONCRETE BRIDGES   129

7.1 Slab Bridges / 129
7.2 Deck-Girder Bridges / 129
    7.2.1 Reinforced Concrete T-Beams / 130
7.3 Box-Girder Bridges / 131
    7.3.1 Box Girder Design / 131
7.4 Prestressed Concrete Segmental Bridges / 132
    7.4.1 Cast-In-Place Balanced Cantilever / 132
    7.4.2 Precast Balanced Cantilever / 133
    7.4.3 Span-By-Span Construction / 133
    7.4.4 Progressive Placement Construction / 135
    7.4.5 Incremental Launching or Push-Out Construction / 135

7.4.6 Range of Application of Bridge Type by Span Lengths Considering Segmental Construction / 135
7.5 Reinforced Concrete Trusses / 135
7.6 Frame Bridges / 137
7.7 Arches / 138
　7.7.1 Introduction / 138
　7.7.2 Structural System / 138
　7.7.3 Abutment of the Arch / 139
　7.7.4 Geometry of the Arch / 140
　7.7.5 Robert Maillart's Bridges / 140
　7.7.6 Concrete Suspension Bridges / 140
7.8 Concrete Cable-Stayed Bridges / 141
7.9 Prestressed Concrete Bridges / 145
　7.9.1 Prestressing Methods / 145
　7.9.2 Economy / 148
　7.9.3 Precast-Beam Bridges / 148
　7.9.4 Cast-In-Place Prestressed Concrete / 148
　7.9.5 The Cantilever System / 148
　References / 149

8. SUBSTRUCTURE—PIERS　　151

8.1 Piers / 151
8.2 Piers with Ice-Breaking Cutwater / 152
8.3 Materials and Construction / 153
8.4 Types of Piers / 155
　References / 158

9. SUBSTRUCTURE—ABUTMENTS　　159

9.1 Introduction / 159
9.2 Types of Abutments / 159
　9.2.1 Wing-Type Abutment / 159
　9.2.2 Straight-Wing Abutment / 160
　9.2.3 U-Type Abutment / 161
　9.2.4 Box-Type Abutment / 161
　9.2.5 Flanking-Span Abutment / 162
　9.2.6 Floating Abutments / 163
9.3 Material / 163

9.4 Design Analysis / 164
   References / 164

10. AESTHETICS IN BRIDGE DESIGN 166

10.1 Introduction / 166
10.2 Requirements for Bridge Aesthetics / 167
      10.2.1 Conformity with Environment / 167
      10.2.2 Economic Use of Material / 167
      10.2.3 Exhibition of Purpose and Construction / 167
      10.2.4 Pleasing Outline and Proportions / 167
      10.2.5 Appropriate but Limited Use of Ornament / 168
      10.2.6 Expressiveness / 168
      10.2.7 Symmetry and Simplicity / 168
      10.2.8 Harmony and Contrast / 169
      10.2.9 Material and Colors / 169
      10.2.10 Proportion / 169
10.3 Causes of Lack of Aesthetics / 170
10.4 Aesthetics of Ordinary Steel Bridges / 170
      10.4.1 Beam Bridges / 171
      10.4.2 Truss Bridges / 171
      10.4.3 Movable Bridges / 171
      10.4.4 Cantilever Bridges / 171
      10.4.5 Arches / 172
      10.4.6 Suspension Bridges / 172
      10.4.7 Cable-Stayed Bridges / 173
      10.4.8 Reinforced Concrete Bridges / 175
  References / 175

11. SPECIFICATIONS AND CODES 177

11.1 General Data / 177
11.2 Loads on Bridges / 179
      11.2.1 Dead Loads / 179
      11.2.2 Live Loads / 180
      11.2.3 Reduction in Load Intensity / 181
      11.2.4 Sidewalk Loading / 183
      11.2.5 Impact / 183
      11.2.6 Longitudinal Forces / 184
      11.2.7 Wind Loads / 184

11.2.8 Thermal Forces / 186
   11.2.9 Uplift / 186
11.3 Distribution of Loads to Stringers, Longitudinal Beams, and Floor Beams / 187
   11.3.1 Position of Loads for Shear / 187
   11.3.2 Bending Moments in Stringers and Longitudinal Beams / 187
11.4 Substructure / 188
   11.4.1 Forces from Stream Current, Floating Ice and Drift / 188
11.5 Earth Pressure / 190
11.6 Seismic Design / 190
   11.6.1 Flow Charts and Examples for Use of Standards / 191
   11.6.2 Applicability of Standards / 191
   11.6.3 Section 3. General Requirements / 191
   11.6.4 Section 4. Analysis and Design Requirements / 191
   11.6.5 Section 5. Analysis Methods / 194
   11.6.6 Section 6. Foundation and Abutment Design Requirements / 194
   11.6.7 Section 7. Structural Steel / 194
   11.6.8 Section 8. Reinforced Concrete / 194
   References / 196

12. METHODOLOGICAL TRENDS IN DESIGN OF BRIDGES   198

12.1 Characteristics of Basic Trends in the Design of Bridges / 198
   12.1.1 Rational Computation Trend / 199
   12.1.2 Creative Trend / 200
   12.1.3 Practical Trends / 201
12.2 Basic Assumptions of Design / 201
   12.2.1 Basic Requirements of the Bridge Under Design / 204
   12.2.2 Additional Requirements of the Bridge Under Design / 206
12.3 Basic Parameters of the Bridge / 210
   12.3.1 Bridge System / 212
   12.3.2 Size of Separate Spans / 213

12.3.3 Type of Span Construction / 213
12.3.4 Types of Supports / 214
12.4 Theoretical Basic Methods of Preliminary Design / 214
12.4.1 Introduction / 214
12.4.2 Practical Methods of Preliminary Design / 218
12.4.3 Choice of Final Alternative of Reinforced Concrete Bridge / 219
12.4.4 Conclusions / 222
References / 223

13. METHODOLOGY OF PRELIMINARY DESIGN   224

13.1 Introduction / 224
13.2 General Consideration for Design of Structural Bridge Scheme / 226
13.3 Sequence of the Work During Design of Bridge Alternatives / 227
13.4 Local Conditions and Solution of General Problem of Construction / 228
13.4.1 Clearance Under Bridges and Underpasses / 231
13.5 Systems of Reinforced Concrete Bridges / 232
13.6 Steel and Composite Bridges / 237
13.7 Supports for Girder Bridges / 246
13.8 Design of the Spans / 249
13.9 Determination of the Amount of Basic Work During the Design of Bridge Alternatives / 253
13.10 Expenditure of Material in Reinforced Concrete, Precast Simple-Span Bridges up to 130 ft / 253
13.11 Weight of Steel in Steel and Composite Span Bridges / 256
References / 259

14. COMPARISON OF ALTERNATIVES   261

14.1 Comparison of Alternatives by Calculated Cost / 261
14.2 Comparison of Alternatives by the Cost of Basic Construction Materials / 261
14.3 Comparison of Alternatives by Conditions of Fabrication and Erection / 262
14.4 Comparison of Bridges Alternatives by Conditions of Performance / 264

14.5 Comparison of Alternatives by External View / 264
14.6 Examples of Bridge Alternatives / 264
    14.6.1 Example 1. Alternatives for a Reinforced Concrete Bridge with a Clear Shipping Span of 98 ft / 265
    14.6.2 Example 1. Comparison of Alternatives of Bridge Structures / 269
    14.6.3 Example 2. Comparison of Bridge Alternatives Across a Large River with a Navigable Clearance Crossed by a Steel Span (Figs. 14.3 and 14.4) / 270
    14.6.4 Example 2. Comparison of Alternatives / 282
    14.6.5 Example 3. Design of Bridge Scheme Across a Large River / 283
References / 295

15. COMPUTER-AIDED DESIGN OF BRIDGES   296

15.1 Preparation of the General Data / 296
15.2 Planning and Bridge Design / 297
15.3 Computer Application / 298
15.4 Bradd-2 System / 299
    15.4.1 Function / 299
    15.4.2 Hardware / 300
    15.4.3 System Software / 300
    15.4.4 Operation / 300
    15.4.5 Output / 301
15.5 New Image System / 301
    15.5.1 Function / 301
    15.5.2 Hardware / 301
    15.5.3 Operation / 301
    15.5.4 Output / 301
15.6 BDES System / 302
15.7 BDS—Bridge Design System / 303
15.8 Geomath System / 304
15.9 Applications of Microcomputers in Bridge Engineering / 304
15.10 DCA Structural Engineering Software / 305
    15.10.1 Structural Designer / 305
    15.10.2 Steel Detailer / 306

15.11 Direct Optimal Design of Continuous Composite Steel Bridge Girder / 307
15.12 Model for the Integration of Design and Drafting Software / 308
15.13 The Effective Use of CADD in Bridge Design / 308
References / 309

INDEX 311

PREFACE

This text discusses the basic principles of bridge design. The book is intended primarily for all engineers, especially the younger ones, who are engaged either directly or indirectly in the design and building of bridges.

Generally, the ability to design bridges consists of a knowledge of the methods of analysis, good judgement, and experience. The history of bridge engineering indicates that bridge design is an area of creative activity for bridge engineers. Therefore, it is necessary to consider certain factors which define this activity. One such important factor is knowledge of the theory of bridge design. At the beginning, designers of bridges were guided by common sense and experience. Further developments were enhanced by knowledge of the theory of structures and properties of the materials. Consideration of different factors and their interconnection permits for each historical period to indicate basic conceptions in the bridge design and its principles.

This text is an attempt to summarize the main technical data which constitute bridge engineering and to indicate steps for designing and comparing alternative solutions. It is also intended to provide bridge engineers and postgraduate students with criteria and methods for preliminary design.

The material contained in this book is presented in 15 chapters. Chapter 1 presents a comprehensive review of the history of bridges, focusing on the most important periods and bridges which made history in bridge engineering. Chapter 2 discusses the choice of bridge location and Chapter 3 treats the design of bridge layout. Chapter 4 investigates the methods of designing bridges that span rivers. In Chapter 5 we analyze structural bridge system and in Chapter 6 we look at typical steel superstructures such as plate and box girders, trusses, arches, and suspension or cable stayed bridges. Chapter 7 discusses typical concrete superstructures. Substructures such as piers and abutments are discussed in Chapters 8 and 9, respectively. In Chapter 10 we focus on the aesthetics of bridge design. Chapter 11 presents the specifications and codes used in the design of bridges and Chapter 12 discusses the characteristics of basic trends in bridge design. The methodology of preliminary design is analyzed in Chapter 13. Chapter 14 considers methods for comparing and choosing among alternative bridge designs. Lastly, in Chapter 15 we discuss computerized planning and design of bridges. Since many variables affect the material presented in this book, the author would appreciate having any errors called to his attention.

M. S. TROITSKY

Montreal, Canada
March 1994

CHAPTER 1

HISTORY OF BRIDGES

1.1 INTRODUCTION

When it is necessary to build a bridge, the question arises: What kind of bridge is it necessary to build? From a design standpoint, there may be many possibilities. Thus the creative capability of the designer plays a large role in answering the question posed above.

The creativity of the bridge designer must of course be grounded in the discipline of engineering. It is also necessary to have a technical mastery of the materials used to build bridges before the design process can begin.

It is also important for the designer to collect and critically analyze data about bridges being constructed worldwide and to apply the results of his analysis to his own creativity. By fully developing his creative capability, and learning from his own work and that of others, the engineer can attempt to perfect the methods of bridge building, thus advancing the art of bridge engineering.

In this book we consider the role such critical analysis has played in the history of bridge building. Throughout history each important structural problem has been solved by producing a few alternatives and undertaking an investigation to arrive at a solution. Likewise, it is necessary to choose the proper bridge system for each particular case by analyzing several preliminary alternatives.

Knowledge of bridge engineering from world practice has great value. For this reason we begin our discussion from a historical perspective. Of course, it is impossible to show all the bridges ever constructed. We chose those that illustrated the most important engineering and design developments.

1.2 DEVELOPMENT OF BRIDGE TYPES

It is said that the history of bridges is the history of civilization. However, achieving progress in bridge engineering was not an easy task. Bridges, as most other engineering structures, began with the "cut and try" process. Some less kind people say the "try and fail" process.

The pioneers used empirical methods. They made some intelligent guesses as to the strength required and built the bridge accordingly. Many centuries passed before man created the five basic types of bridges: the beam, the cantilever, the arch, the suspension, and the truss.[1-16] The first four types were copied from nature long before recorded history began.

The natural example of the simple beam bridge is that of a fallen tree spanning a stream (Fig. 1.1). The next step was to use a stone slab as a bridge (Fig. 1.2).

Quite probably, primitive man discovered the principle of the cantilever bridge at a very early stage of bridge development. He made use of a cantilever to construct longer spans than he was able to build with simple beams. Timber beams or stone slabs projecting out one above the other represented such bridges (Fig. 1.3).

Natural bridges of stone have also been formed, where the action of water has worn away rock until only an arch was left, high above the river bed (Fig. 1.4).

The suspension or cable bridge is illustrated in nature by the swinging vine, utilized by animals and people to pass from one tree to another over a stream (Figs. 1.5 and 1.6). In its simplest form a suspension bridge consists only of cables and unstiffened roadway. Many primitive bridges of this kind were built

Figure 1.1 Fallen tree as a bridge.

1.2 DEVELOPMENT OF BRIDGE TYPES

Figure 1.2 Stone slab as a bridge.

as shown by Figures 1.7 and 1.8. It is obvious that such an arrangement is too flexible for safety and comfort.

In primitive suspension bridges, the roadway was often laid on top of the cables. But this position was inconvenient, and bridge builders discovered that a level roadway could be obtained by suspending the roadway from the iron chain cable (Fig. 1.9). The first suspension bridge using this system was erected

Figure 1.3 Cantilever bridge.

4 HISTORY OF BRIDGES

Figure 1.4 Natural arch bridge.

in Italy in the sixteenth century (Fig. 1.10). Since the beginning of the nineteenth century, flat iron bars were used for cables. Finally, the truss-type bridge belongs almost exclusively to modern civilization.

In the fifteenth century Leonardo da Vinci was the first to investigate the strength of beams and the forces in triangular structures. Figure 1.11 illustrates his design for a timber truss bridge.[18] Historically, bridges may be conveniently assigned to the five periods discussed below.

Figure 1.5 Swinging vine as suspension bridge.

Figure 1.6 Monkey bridge. Reprinted from Waddell, J. A. L., *Bridge Engineering*, Vol. I, Wiley, New York, 1916.

1.3 THE ANCIENT PERIOD

Today bridge engineering is considered a science. However, about 100 years ago it was hardly worthy to be termed an art, and 150 years ago it was no better than a trade. But while bridge building as a learned profession is thus of relatively recent origin, it must not be thought that previous centuries made no contributions to our knowledge of bridge construction.

Primitive man must have built many crossings over shallow streams by piling rocks for piers and covering them with slabs of stone, logs, or falling trees so as to span small rivers. Suspension bridges across the water were built from the overhanging branches of opposite trees. From such data it may be seen that the evolution of bridge engineering resulted from the evolution of the form of structure, the materials of construction, and the methods of design, fabrication, and erection.

6 HISTORY OF BRIDGES

Figure 1.7 Primitive suspension bridge, South America.

Figure 1.8 Primitive suspension bridge over the Pampas River, South America.

1.3 THE ANCIENT PERIOD   7

Figure 1.9 Early suspension bridge.

The earliest bridge of which there is any authentic record was built over the Euphrates at Babylon about 780 B.C. Herodotus described it, writing in 484 B.C. It was a short-span structure, 35 feet wide, of timber beams resting on stone piers.

Another early form of construction used by primitive man was the suspension type. It was used in remote ages in China, Japan, India, and Tibet. It was

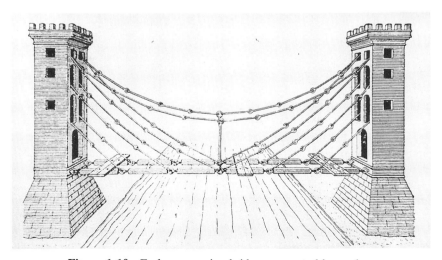

Figure 1.10 Early suspension bridge, supported by eyebars.

Figure 1.11 Leonardo da Vinci's timber truss bridge.

used by the Aztecs of Mexico and the natives of Peru and other parts of South America. The cables of these primitive structures were made of vines twisted onto straps of hide and fastened to trees or other permanent objects on shore.

One suspension bridge of 330-ft span is said to have been built in China about 65 A.D., and it is believed that others had been completed in that country many centuries earlier. Iron chains for suspension cables were adopted in both India and Japan more than 500 years ago, while rope was employed for the same purpose in Europe, India, and South America several centuries back.

Another early form of bridge was the cantilever span. The Chinese are believed to have constructed cantilever bridges many centuries ago. As far back as 1100 B.C. it is known that the ancient Greeks employed the cantilever.

A later form of bridge was the arch. The construction of masonry arches began before the days of recorded history. Corbeled stone arches were used by the Egyptians in the Pyramid of Gizeh, dating back some 3000–4000 years before the Christian Era. Brick arches of crude form were found in the ruins of Thebes in structures that were probably built about 2900 B.C. The inhabitants of the valleys of the Euphrates and the Tigris also were familiar with the arch at a very early period. The Babylonians built pointed brick arches certainly as early as 1300 B.C. About 2000 B.C., the River Euphrates in the City of Babylon was crossed by a single brick arch 30 ft wide and 660 ft long. The Chinese have employed the true semicircular arch for ages, although their old spans were always short. Finally, the Caravan Bridge over the River Meles, at Smyrna in Asia Minor, is of a very early though unknown date, and is believed by many to be the oldest existing bridge. It is a single span, 40 ft in length, and is still in use. It must be about 3000 years old, and most of it is still in its original condition.

1.4 THE ROMAN PERIOD

The Roman Period dates from 300 B.C. and covers a period of about 600 years. The Romans were the first real bridge engineers. They built bridges in wood, stone, and concrete. The Romans always built their stone and concrete bridges with arches. They solved the complicated engineering problems of how to rest their massive spans on underwater piers and how to protect the piers from floods. Today, Roman arches still stand in Italy, Spain, and France as monuments to their genius.

The oldest Roman bridge, according to history, was the Pons Sublicius,[19] named for the sublicae, or wooden beams, from which it was built across the Tiber River in 620 B.C. (Fig. 1.12). The most celebrated of all the early bridges was Caesar's pile trestle,[20] built in 10 days over the Rhine River, during 55 B.C. (Fig. 1.13). Caesar's Rhine bridge was founded on groups of wooden piles. A series of these piles and cross beams were carried right across the river, and then logs were laid along them to form the roadbed of the bridge.

The Romans' greatest undertaking was probably a large timber arch bridge over the Danube River, which was constructed by order of the Emperor Trajan[21] in 104 A.D. It contained 20 wooden arch spans resting on cut-stone piers (Fig. 1.14). There is good reason to believe that the length of each span may have been as much as 170 ft.

To erect underwater piers, the Romans made use of cofferdams. They drove the piles into the riverbed around the intended site of their piers, and lined the rams of piles with clay to make them watertight. Then the interior of the cofferdam could be pumped out and concrete poured in to form the pier. Where the riverbed was too deep for workmen, the solution was to drop concrete blocks to the bottom to create an artificial floor. The bridge was 150 ft high, 60 ft wide, and 4500 ft long. It lasted only 30 years and the Romans destroyed it themselves during the war.

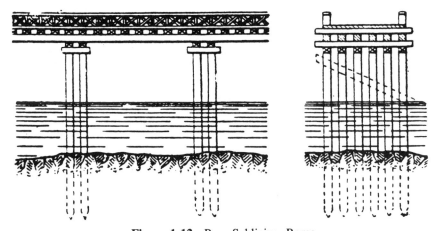

Figure 1.12 Pons Sublicius, Rome.

Figure 1.13 Caesar's bridge.

Figure 1.14 Trajan's bridge.

1.4 THE ROMAN PERIOD

Some of the greatest Roman bridges were aqueducts designed to carry not people but a water supply. The most well-known existing Roman aqueduct is the Pont du Gard, near Nimes in France[22] (Fig. 1.15). Stretching across the valley of the River Gard for 860 ft, the Pont du Gard is a three-tiered bridge that reaches a height of 155 ft above the river. The lowest tier consists of six large arches, 31–80 ft in span. The topmost tier carriers the water channel.

The aqueduct of Segovia in Spain[23] has two stories and is 119 ft high at the center. The material is cut stone put together without mortar, and the total length of the structure is 2700 ft (Fig. 1.16).

Two-thousand years ago the Romans built eight stone bridges across the Tiber, and six of these are still standing today. The most celebrated is the Bridge Sant'Angelo[24] (Fig. 1.17). Its roadway, 25 ft wide, rests atop seven arches, the longest one spanning 60 ft.

The Romans abandoned timber bridges in favor of stone bridges, which required that the arch be perfected. Following the lead of the Etruscans, the Romans used the strong semicircular arch. An arch exerts not only a downward thrust but an outward one, having a tendency to spread or thrust. In order to hold such an arch together, the Romans were forced to use heavy abutments or buttresses. The Romans also used concrete consisting of pozzolana, a red volcanic earth, mixed in a kiln with limestone. This concrete set quickly even under water. To make arches, the Roman engineers would build a centering

Figure 1.15 Pont du Gard Aqueduct, France.

12 HISTORY OF BRIDGES

Figure 1.16 Segovia Aqueduct, Spain.

Figure 1.17 Bridge Sant'Angelo, Rome, Italy.

of wood or brick to hold the sections of the arch in place until the entire structure was complete.

Looking back nearly 2000 years, in light of our modern knowledge of bridge construction, it is quite surprising to note how little one finds to criticize in the work of the ancient Roman builders. The Romans built exclusively semicircular arches. The Roman engineers, despite their skill in using cofferdams, did not succeed completely in mastering the bottom of the river. However, the Romans built bridges of monumental strength and striking beauty. Their engineering triumphs long outlasted their own empires.

1.5 THE MIDDLE AGES

The Middle Ages in Europe, from the eleventh to the sixteenth centuries after Rome fell, comprise the third period in the history of bridge construction.

In the twelfth century Peter of Colechurch began work on the Old London Bridge.[25] The job took 33 years. London Bridge did not differ in any essential way from a typical Roman bridge. Stone piers supported arches of unequal spans. However, instead of a Roman semicircular arch, the gothic arch was used, which was easier to construct because it placed less load on the wooden centering while being built. The spans of the arches ranged from 15 to 34 ft, and the piers were 18 to 26 ft in width. The river had to flow through 19 small and unequal openings (Fig. 1.18). To pay the cost of maintenance, the bridge

Figure 1.18 View of the Old London Bridge.

authorities decided to build houses and shops along the roadway on the bridge. Despite the famous song, London Bridge never did fall down—it stood for 600 years. In 1831, when a new London Bridge was completed, the remains of Peter of Colechurch's Bridge were cleared from the river.

The bridge at Avignon over the Rhone River[26] in Southern France was built in the twelfth century under the direction of Saint Benezet, who was once a shepherd boy. It originally contained 22 spans, the longest 110 ft, and the bridge was nearly 3000 ft long. The roadway of the bridge was a shape that would have mystified today's engineers. It was 16 ft at its widest point. On the Avignon side, it squeezed down to only $6\frac{1}{2}$ ft; at this point Benezet built a chapel.

The bridge was bowed upstream to better resist the force of the current. The slender arches were rounded, their curve was elliptical. The piers have a triangular shape or "cutwaters." This is the principle we now call "streamlining." Four arches still remain today (Fig. 1.19).

The Immortal, the Ponte Vecchio bridge[27] in Florence, Italy, built in the fourteenth century, is another old stone bridge which stands to this day (Fig. 1.20). It crosses the River Arno in flat, segmented arches. The symmetrical Ponte Vecchio has a central span of 100 ft and a pair of 90-ft side spans. Above the roadway rose a two-story arcade, the upper gallery connecting two famous palaces, the lower lined with a double row of jeweler's shops. This bridge was completed in 1367.

Figure 1.19 The bridge at Avignon, France.

Figure 1.20 The Ponte Vecchio bridge, Italy.

Work on the Pont Neuf bridge[28] started in 1578 between the downstream end of the Ile de la Cote and both banks of the Seine. This was the Pont Neuf, or "the new bridge," which is now the oldest surviving bridge in Paris. The Pont Neuf reaches in two directions, a long arm of seven arches going to the right bank and a short arm of five arches to the left (Fig. 1.21). The bridge was opened in 1607.

The Rialto,[29] probably the best known bridge in Europe, was built in Venice over the Grand Canal in the sixteenth century. The bridge has an extreme length of 158 ft, open span of 85 ft, and width of 72 ft. On the roadway are two rows of shops with a passageway between them. The footwalks on the outside are carried on projecting brackets (Fig. 1.22).

The builder faced the problem of founding a heavy stone bridge on soft subsoil. His idea was to drive 6000 piles into the mud on either side of the canal. They were driven in such tight clusters that they practically touched each other, forming a nearly solid wooden base. The piles were cut level and capped by three layers of timber fastened by iron clamps. Layers of stone were placed above these timbers and the arch itself.

The Rialto Bridge is considered not only an outstanding engineering accomplishment, but an artistic triumph, for it is one of the world's most beautiful bridges.

Figure 1.21 The Pont Neuf, France.

Figure 1.22 The Rialto bridge, Venice.

1.6 AN AGE OF IRON AND STEEL BRIDGES

With the Industrial Revolution a new era of iron and steel was born and the seven bridges described below made history.

When iron was first used in bridge building, the builders imitated stone construction. Consequently, the first iron bridge, the Coalbrookdale bridge over the Severn River in England,[30] was built of separate wedge-shaped pieces that formed the arch ring. The world's first iron bridge, it was built of cast iron in 1776, with a span of 100 ft. It is an arch bridge made of five semicircular cast-iron ribs rising 45 ft from the banks of the Severn River to carry the roadbed (Fig. 1.23). The pioneering Coalbrookdale bridge still stands. In recent years its abutments have had to be strengthened, but the ironwork itself remains as it was built.

In their search for a design sufficiently rigid for railway use, some engineers built tubular bridges. The most outstanding among these is the old Britannia bridge[31] over Menai Straits, England. The Britannia bridge is constructed in the form of two rectangular tubes, built of wrought-iron plates. It carries two tracks of the railway (Fig. 1.24).

The bridge was designed by the famous engineer Robert Stephenson and, completed in 1850, is still in regular use. It consists of two main spans of 460 ft each, two side spans, and two approach spans. Each tube is 1511 ft long, and one tube of the central span weights 1600 tons. The shore spans were built on falsework, but the center ones were floated into place on pontoons and lifted with hydraulic jacks. Some doubts regarding the strength of the bridge under the vibration of trains inspired a fear that suspension chains might be neces-

Figure 1.23 Coalbrookdale Bridge, England.

Figure 1.24 The Britannia bridge, England.

sary. To allow for these, the towers were extended considerably above the track level for support of a suspension system. After completion of the bridge, its designers found their work sufficiently strong and rigid without the suspension chains. Despite the greatly increased weight of modern trains, the bridge continues in use to this day without any suspension chains.

After the erection of Britannia bridge, the old Victoria bridge,[32] the largest tubular bridge, was built in 1860 over the St. Lawrence River at Montreal. (Fig. 1.25). To meet the needs of increased traffic, it was replaced by the truss type in 1898.

The earliest steel arch bridge was built in the United States across the Mississippi River in St. Louis by James Eads.[33] The Eads bridge is a monument to the originality and genius of this great engineer—a pioneer master builder. People said his dream was crazy. Engineers declared the project to be dangerous and bold. But the bridge, completed in 1874, still stands. It carries heavy railway and highway traffic.

The bridge is remarkable for its beauty as well. Eads made the center span somewhat longer than the two side spans, and so achieved a nicely balanced composition. The length of the center span is 520 ft, flanked by two 502-ft sidespans (Fig. 1.26). The midwater piers were built on caissons sunk deeper than had ever been done before, to the rock at a depth of 128 ft.

Eads erected arches by tying them with steel cables to the piers. Each of the three arch spans consisted of parallel arched tubes linked by diagonal bracing. When the first arch was nearly complete, the two halves of the arch were built toward each other and a gap of only a few inches separated them. However, the heat of the sun had so expanded the steel tubes that the gap left for

1.6 AN AGE OF IRON AND STEEL BRIDGES 19

Figure 1.25 The Victoria bridge, Montreal, Canada. Reprinted from Legge, C. A., *A Glance at the Victoria Bridge, and Men Who Built It*. Printed and published by John Lovell, Montreal, 1860.

the last section was too small to receive it. To wait for cold weather was impossible, for the contract called for completion of the arch by a specified date. To shrink the arch, the men packed 45 tons of ice around it and the arch was safely closed. The project was saved.

St. Louis bridge was the biggest steel bridge that had ever been built at that time. Eads' use of pneumatic caissons for the foundations was another great

Figure 1.26 The Eads bridge, St. Louis, USA.

20 HISTORY OF BRIDGES

accomplishment. This bridge transformed St. Louis into an important rail center.

The greatest of all cantilever bridges is the world-famous Firth of Forth bridge[34] in Scotland. Although completed in 1889, it is still in use. The bridge carries two lines of rail track. This cantilever bridge has two anchor arms, three towers, and two 1700-ft cantilever spans. Each cantilever span is composed of two cantilever arms and a suspended span. The total length, including approaches, is 8300 ft (Fig. 1.27). Most of the compression members are hollow tubes, the largest being 12 ft in diameter. The towers themselves consist of four steel tubular members. The tubes are not perfectly upright for they lean toward each other across the deck of the bridge, so that a section of the structure is like a triangle with the top cut off. The towers are nearly 350 ft high.

When the time came to close the first span, a problem developed opposite to that of the St. Louis bridge. Unusually cold weather caused the metal to contract so the arms would not meet, and a fire had to be lit along the steelwork so that the last section could be linked.

The Forth bridge was an engineering success. It established the cantilever bridge as a rival of the suspension bridge for long-span crossings. Some engineers, looking into the giant trusses of the cantilever form, regarded it as a safer and more stable structure than a suspension bridge.

The design for the Quebec cantilever bridge[35] called for a central span of 1800 ft from center to center of main piers. In 1904 work began on a bridge across the St. Lawrence River in Quebec. In 1907, the shoreward and river-

Figure 1.27 The Firth of Forth bridge, Scotland.

1.6 AN AGE OF IRON AND STEEL BRIDGES

ward spans of the south cantilever were finished. One day it was discovered that the incomplete suspended span had dipped a fraction of an inch toward the water; an entire south cantilever arm ripped loose and into the St. Lawrence went 9000 tons of steel and 86 workmen. The Quebec authorities ordered a new bridge—again of the cantilever type. In 1914 the new foundations were in place, and within two years the cantilever arms were complete. In 1916 barges floated the 640-ft long suspended span. As it rose, one of the links snapped and the 5000-ton span plunged into the river. After one year a new span was successfully put in place and the bridge opened in 1918 (Fig. 1.28).

The Hell's Gate arch bridge[36] over the East River in New York is not only one of the heaviest and longest steel arches ever built, it is also one of the most beautiful steel bridges ever constructed (Fig. 1.29). The big steel arch with its massive abutments and its gracefully sweeping lines, carrying trucks 140 ft above the river, forms a striking picture. Its span of $977\frac{1}{2}$ ft made it the longest steel arch in the world from its completion in 1917 until the Bayonne bridge and the Sydney Harbour bridge were opened in 1931 and 1932. The striking curves of the Hell Gate arch did not result entirely from artistic constructions; they were largely necessitated by the rigidity and clearance requirement of a structure designed to carry heavy steel trains hauled by powerful locomotives.

In 1931 the Bayonne bridge[37] was opened for traffic over Kill Van Kull. It is the longest steel arch bridge in the world. The span is 1652 ft between the hinges, supporting a roadway at a clear height of 150 ft above the water at the

Figure 1.28 The Quebec cantilever bridge, Canada.

22 HISTORY OF BRIDGES

Figure 1.29 The Hell's Gate arch bridge, New York.

center of the span (Fig. 1.30). The Bayonne bridge carries a 40-ft roadway for a total of four lanes of traffic and it has also a sidewalk.

1.7 AN ERA OF SUSPENSION BRIDGES

The grandfather of all great suspension bridges is the Menai Straits bridge, England,[38] built by Thomas Telford in 1825. Telford's brilliance created a bridge that was 60 years ahead of its time. He designed a suspension bridge

Figure 1.30 The Bayonne bridge, New Jersey.

1.7 AN ERA OF SUSPENSION BRIDGES

with stone towers and chain cables from which the roadway would be hung. The total length of the bridge is 1710 ft. The huge towers are 153 ft high and 580 ft apart. Two sets of chains, one on each side of the tower, hold up a 30-ft wide roadway (Fig. 1.31). The old iron chains remained in service until 1939 when steel bars took their place. This bridge is still standing.

In 1846 the directors of the American and Canadian bridge Companies asked the famous American engineer, John Roebling, to give his opinion concerning a proposed railway bridge that would cross Niagara Falls. When it was announced that Roebling proposed a railway suspension bridge with an 820-ft span across Niagara, engineers proclaimed the project was impossible. However, the work started in 1851. To carry the first wire across the gorge, Roebling offered a prize of $10 to the first boy who would fly a kite with one end of the wire across the Niagara gorge.

The suspended structure had an upper deck for railway and a lower deck for highway traffic. The two decks were connected together by a 20-ft deep and 20-ft wide stiffening truss made of timber. The span was suspended from four cables, each having a diameter of 10 in., supported by four masonry towers, and the ends of the cables were carried to anchorages in the solid rock (Fig. 1.32).

In 1855, Roebling completed this first successful railway suspension bridge in the world.[39] It was the first suspension bridge ever built with stiffening

Figure 1.31 The Menai Strait suspension bridge, England. Reprinted from Tyrrell, H. G., *History of Bridge Engineering*, published by the authors, Chicago, 1911, p. 212.

24 HISTORY OF BRIDGES

Figure 1.32 The Niagara Suspension Bridge, USA.

trusses. This was a major contribution to bridge engineering art—the greatest in many centuries. By providing stiffening trusses, Roebling introduced a high safety measure for the stability of suspension bridges.

For more than 40 years the Niagara Suspension Bridge carried heavier and heavier traffic loadings. Finally, it was $2\frac{1}{2}$ times heavier than the original design loading and the bridge was dismantled.

In 1867 John Robeling was appointed chief engineer of the proposed bridge over the East River in New York. With a prophetic vision of the future development of New York and Brooklyn, he selected a location for the bridge, and in three months completed the plans for it. The proposed span was 1600 ft. Engineers declared Robeling's project as impractical and fantastic! As the actual work was about to start, John Roebling died accidentally and his son, Washington, took over his father's job as chief engineer. In 1883, almost 14 years after it had been started, the bridge was finished.

The Brooklyn Bridge is remarkable not only for its span and early use of pneumatic caissons, but also because it was the first bridge in the world on which steel wire was used instead of iron. The Brooklyn Bridge, with a 1600-ft main span, is suspended by four cables 16 in. in diameter (Fig. 1.33). Each cable is composed of more than 5000 parallel steel wires. The stiffening trusses hang from the cables by wire-rope suspenders at a 7-ft spacing.

A unique condition exists in the Brooklyn Bridge. The steel rollers under the saddles on top of the towers become rusty or "frozen." This resulted in an unbalanced cable pull on the tower tops, and has produced serious bending strains in the masonry towers. However, the towers rest on compressible timber caissons. The elastic compressibility of the timber permits the towers to rock on their base, therefore relieving the strains in them.

Figure 1.33 The Brooklyn suspension bridge, New York.

Perhaps the most distinctive feature of the Brooklyn Bridge is the system of inclined stays radiating downward from the towers to the deck. Roebling introduced them for the stability of the span against the wind.

It has been said that the one single factor that contributed more than anything else toward the creation of the great city of New York was the building of the Brooklyn Bridge. This great American bridge[40] is still standing. The essential elements of the structure—the towers, cables, and anchorages—will last perhaps for centuries. John Roebling solved the technical problems that made it possible to build our modern suspension bridges! He predicted that the steel cables could support a span of over 3000 ft. In less than two generations, the great suspension bridges that we know today proved him correct.

The George Washington suspension bridge[41] conception originated with the great engineer Ammann. Construction began in 1927 and the bridge was opened in 1931. This great bridge has a main span of 3500 ft from tower to tower, with end spans of 630 and 610 ft, making the total length of the main bridge 4740 ft (Fig. 1.34).

The finished structure, with the lower deck added later, has a clearance of 213 ft over midriver. Its weight is carried by four 36-in. cables that run from anchorage to anchorage over gigantic towers reaching 635 ft above the water. It has an eight-lane roadway on the upper deck with four lanes on the lower deck.

The steelwork for the towers had been planned only as a reinforcing skeleton to be covered by concrete and granite, but as the steel skeleton soared upward, the natural, functional beauty of the steelwork fascinated everyone and the

Figure 1.34 The George Washington suspension bridge, New York–New Jersey.

steel frame remained uncovered. Each of the four cables is a yard in diameter and nearly a mile long.

The famous Golden Gate suspension bridge,[42] built in 1937, has a 4200-ft main span and its two end spans measure 1125 ft each, so that the total length of suspended structure is thus 6450 ft (Fig. 1.35). Its great towers reach a height of 746 ft and they rest upon piers sunk down to solid rock 100 ft below water. To eliminate the possibility of damage to the bridge resulting from an earthquake, the foundation has been sunk 25 ft into the rock. The total width of the bridge floor is 81 ft, including a 60-ft roadway and two $10\frac{1}{2}$ ft sidewalks. Two cables, each $36\frac{1}{2}$ in. in diameter, carry the structure at a height which leaves a 220-ft clearance over the water at the middle of the span. Its San Francisco pier was the first bridge pier ever built in deep open water and the bridge itself is the only structure ever built across the outer mouth of an important ocean harbor.

The Mackinac Straits suspension bridge[43] was completed in 1957 with a 3800-ft center span. Besides this tower-to-tower span, the cables run to their anchorages, for an overall cable length of 8344 ft. The towers for the suspension span rose 552 ft above the strait. The designer, Steinman, braced his bridge against the winds of the strait with trusses 38 ft high (Fig. 1.36). Certain novel features in the placement of these trusses were intended to give additional protection against the wind. To guard against the danger of ice flows, the designer put down a mammoth foundation for his piers.

The Verrazzano-Narrows suspension bridge[44]—the largest suspension bridge built in North America, having a main span of 4260 ft—was opened for traffic

1.7 AN ERA OF SUSPENSION BRIDGES   27

Figure 1.35 The Golden Gate suspension bridge, California.

Figure 1.36 The Mackinac Strait suspension bridge, Michigan.

28 HISTORY OF BRIDGES

Figure 1.37 The Verrazzano-Narrows suspension bridge, New York.

in 1964. The Verrazzano-Narrows spans the entrance to New York Harbor. Its 700-ft towers are as tall as 70-story building. They rise so high that the Earth's curvature had to be taken into account. In fact, the towers are about $1\frac{3}{4}$ in. farther apart at the top than at the bottom. The four cables, each 3 ft in diameter, cost more than the entire Golden Gate Bridge (Fig. 1.37). About 150,000 miles of wire were required, enough to encircle the earth nearly six times. A typical cross-section of the Verrazzano-Narrows bridge shows a double deck; it carries 12 traffic lanes (Fig. 1.38).

The Humber suspension bridge[45] over Severn River in England, finished in 1981, has a 4626-ft main span and is the longest suspension bridge in the world

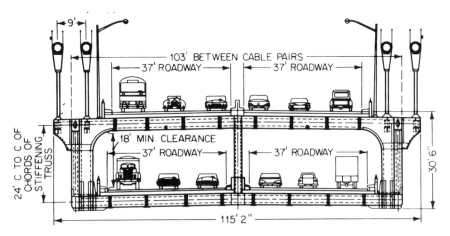

Figure 1.38 Verrazzano-Narrows suspension bridge cross-section.

Figure 1.39 The Humber suspension bridge, England—longest bridge span in the world.

(Fig. 1.39). The Humber suspension bridge is much more slender than the Verrazzano-Narrows, having four traffic lanes instead of 12. An unusual deck using a six-sided box girder instead of conventional stiffening trusses and suspending inclined hangers reduce both the weight and the cost of this bridge (Fig. 1.40). The towers are 533 ft $1\frac{5}{8}$ in. tall from the water level and are $1\frac{3}{8}$ in. out of parallel, to allow for the curvature of the earth. Including side spans, the bridge stretches 1.37 miles.

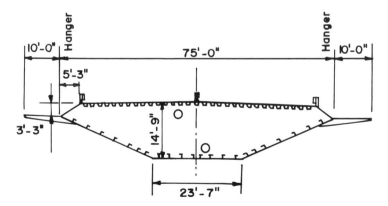

Figure 1.40 The Humber suspension bridge cross-section.

1.8 AN ERA OF CABLE-STAYED BRIDGES

During the past three decades cable-stayed bridges found wide application especially in Western Europe and other parts of the world. The successful application of the cable-stayed system was realized with the introduction of high-strength steel and orthotropic-type decks, the development of welding techniques, and progress in structural analysis. The development and application of electronic computers opened up new and practically unlimited possibilities for the exact solution of these statically indeterminate systems, and for precise statical analysis of their three-dimensional performance.

A comparison between modern types of suspension bridges and the cable-stayed bridge indicates that the cable-stayed bridge is superior to the suspension bridge. The superiority of the cable-stayed bridge may be based on a comparison of their structural characteristics following an analysis proposed and developed by Gimsing. When considering the ratio of pylon height and length of main span, it was found that cable-stayed bridges cost less. Considering deflection for both symmetrical and asymmetrical traffic loading over half of the length of the main span, the suspension bridge has a greater deflection at midspan than does the cable-stayed bridge.

The outstanding advantage of the cable-stayed bridge is that it does not require large and heavy anchorages for the cables as do suspension bridges. The anchor forces at the ends of the cable-stayed bridge act vertically and can usually be balanced by the weight of the pier and its foundation without much additional cost. The horizontal component of cable force is taken by the superstructure girder in compression or tension. Existing cable-stayed bridges of steel, concrete, and composite provide useful information regarding this new system and some typical examples of these bridges are described below.

The Saint-Nazaire cable-stayed bridge,[46] completed in 1974 over the Loire River in France, is currently one of the largest cable-stayed bridges in the world, with a main span of 1325 ft (Fig. 1.41). The total length of this cable-stayed steel structure is 2361 ft and is divided into three spans supported by stay cables. The steel deck of the bridge has the streamline shape of a box-section girder 43 ft wide at the top. The towers set up on the main piers appear as A-shapes having a height of 223 ft above the piers. The stay cables are arranged on sloping planes and vary in diameter from $2\frac{4}{5}$ to $4\frac{1}{8}$ in. according to their location in the superstructure.

The cable-stayed Maracaibo bridge[47] designed by Morandi was completed in 1962 in Venezuela. The Maracaibo bridge differed in many aspects from previous cable-stayed bridges. First, both the pylons and the stiffening girder were made of concrete, thereby introducing a material that had not before been used in the main elements of cable-stayed bridge structures. Furthermore, it was the first multispan cable-stayed bridge (Fig. 1.42).

It includes five 775-ft main spans with a vertical clearance of 148 ft above the lake's navigation channels and supported by inclined cables. Each 775-ft long span consists of a 2 ft × 311 ft cast-in-place deck at both ends and a

1.8 AN ERA OF CABLE-STAYED BRIDGES   31

Figure 1.41 Saint-Nazaire cable-stayed bridge, France. Reprinted from Troitsky, M. S., *Cable-Stayed Bridge*, 2nd ed., Van Nostrand Reinhold, New York, 1988, p. 55.

Figure 1.42 The Maracaibo concrete cable-stayed bridge, Venezuela.

Figure 1.43 Annacis Island cable-stayed bridge. The longest cable-stayed bridge in Canada.

153-ft suspended prefabricated deck. The cantilever span is supported by inclined cables suspended from the top of a 305-ft four-legged pier of two inclined A-frames linked at the top by a transverse girder. Each suspended span consists of a complex of six beams. The cables have 16 strands, each $2\frac{7}{8}$ in. in diameter, and are protected against corrosion by a galvanization of the wires in the cables. The 57-ft wide bridge deck carries four lanes of traffic with a 4-ft central median and two 3-ft sidewalks.

The Annacis Island cable-stayed bridge,[48] of composite design, crosses the Fraser near Vancouver. Completed in 1987, it was the longest span cable-stayed bridge in the world, measuring 3052 ft, with a central span of 1525 ft between the towers (Fig. 1.43). The width of the bridge is 105 ft, and the deck is designed to carry six lanes of traffic. Precast concrete slabs of the deck span steel floor beams, which are supported by two steel plate girders. The bridge has two linked H-shape towers 506 ft high of reinforced concrete and support fan-type arrangement of the cables.

1.9 AN ERA OF CONCRETE BRIDGES

1.9.1 Reinforced Concrete Bridges

Natural cement was applied to bridge construction in the early part of the nineteenth century. The development of the Portland cement industry, most of which took place after 1855, provided a more reliable material. And as a result, plain concrete came into very extensive use both for arch bridges and the substructures of other forms of bridge construction. The introduction of Portland cement was responsible for the development of a new material known as construction-reinforced concrete. The advent of reinforced concrete has extended the development of beams and arches, bringing into common use the continuous girder and the hingeless arch. The first reinforced concrete bridge was constructed soon after the discovery of the art of making reinforced concrete structures. The early bridges were arch bridges; the first was constructed in France in 1875 with a span of 54 ft, and a width of 13 ft. The construction of reinforced concrete bridges did not really begin, however, before about 1890.

In the 1890s numerous reinforced concrete arch bridges were built. The number of reinforced concrete bridges as well as the span length of the bridges increased enormously during the first decade of the twentieth century. In 1911 the Tiber bridge in Rome was completed, having a free span of the arch ribs of 328 ft and arch rise of only 32.8 ft. In 1923 the span length reached 400 ft in the Coppelen Memorial Bridge, Minneapolis, MN. In 1929 the famous Plongastel bridge over the Elorn River at Brest, France was completed with three spans of 612 ft, and in 1940 the arch bridge at Esla, Spain was completed with a span of 645 ft.

Slab and girder bridges of relatively short span have been built extensively in reinforced concrete during the last decades. The largest span on record for a reinforced concrete girder bridge is 256 ft. The basic systems of reinforced and prestressed concrete bridges are:

1. Slab bridges
    a. Reinforced
    b. Prestressed
2. Deck-girder bridges
    a. Reinforced-concrete T-beams
    b. Prestressed concrete stringers
3. Box-girder bridges
    a. Reinforced
    b. Prestressed
4. Continuous bridges
5. Arch bridges
    a. Open spandrel
    b. Filled spandrel
    c. Tied
6. Rigid-frame bridges
7. Cable-stayed bridges

New bridge forms were introduced by Robert Maillart from 1896 to 1940, the so-called arched discs.

34 HISTORY OF BRIDGES

In 1943 the Sando bridge[49] in Sweden was erected, a reinforced concrete arch type having an 866-ft span, only 8 ft thick at the crown and 14 ft at the ends, which means that the span is only about $L/100$ as thick. Pairs of circular columns support the roadway, which runs over the arch (Fig. 1.44). The arch was built on the wooden trestle propped up by 13 groups of long piles.

The Sydney concrete arch has a span of 1000 ft (Fig. 1.45). This elegant six-lane bridge over the Parramatta River, the western branch of Sydney Harbor, was the world's largest concrete bridge[50] when it was completed in 1964.

1.9.2 Prestressed Concrete Bridges

The modern development of prestressed concrete is attributed to Freyssinet of France, who started using high-strength steel wires for posttensioning and prestressing concrete beams in 1928. In 1940 the Magnel system of posttensioning was developed by Magnel System of Belgium.

The early development of the prestressed concrete industry in the United States and Canada, which started in the 1950s, was oriented toward factory production of precast prestressed elements for highway bridges. The first major prestressed concrete bridge built in the United States was Walnut Lane Bridge in Philadelphia, built in 1951.

A record 787 ft. long prestressed concrete twin box girder for a highway bridge was built in 1976, 150 miles southwest of Tokyo. The bridge crosses the mouth of Hamana-Ko Lane. The total length of the bridge is 2066 ft,

Figure 1.44 The Sando bridge, Sweden.

Figure 1.45 The Sydney arch bridge, Australia.

having a width of 30 ft. The center span box girders are flanked by spans 439 and 180 ft long. The box girders were cantilevered in 10- to 17-ft sections by the large travelers and 8- to 13-ft sections by the small ones. Over the main piers the twin girders are 45 ft deep and have 20-in. thick walls. The depth decreases to 13.5 ft at the center, with 10-in. walls. The girders are continuously prestressed from the expansion joint at the midpoint of the center span to the end piers.

Prestressed concrete segmental bridges began in Western Europe in the 1950s. For a crossing over the Lahn River in Baldwinstein, Germany, Finsterwalder was the first to apply cast-in-place segmental construction to a bridge in 1950. The first application of a segmental bridge in North America was on the Laurentian Autoroute near Ste. Adele, Quebec, in 1964. In 1973 the first U.S. precast segmental bridge was opened to traffic in Corpus Christi, Texas. Up to 1981 in the United States more than 80 segmental bridges were completed.

Prestressed concrete segmental bridges may be identified as precast or cast in place and categorized by method of construction as balanced cantilever, span-by-span, progressive placement, or incremental launching. Prestressed concrete segmental construction extended the practical span of concrete bridges to about 800 ft (250 m) or even 1000 ft (300 m). With cable-stayed bridges, the span range can be extended to 1500 ft.

1.10 PROPOSED FUTURE BRIDGES

1.10.1 Messina Straits Suspension Bridge

The world's longest suspension bridge may link Sicily with mainland Italy by the end of this century.[51] The question is whether the Straits of Messina, which divides Sicily from the mainland, will be spanned by that bridge, by a submerged tunnel anchored to the seabed, or by a tunnel below the seabed. Among the principal challenges for the proposed bridge are the cables for the record-breaking span and the sandy soils in which the bridge anchorages must set.

The projected 10,824 ft (3300 m) span bridge would be twice as long as the Humber bridge in Great Britain, the longest suspension bridge now existing. Its towers would be 1312 ft (400 m) tall (Fig. 1.46). The bridge would carry both rail and highway traffic, as would the tunnel. The Straits pose formidable engineering challenges for the proposed crossing. The area is one of the world's most earthquake-prone regions; winds can reach 100 km/h, and tidal currents run up to six knots.

Figure 1.46 Proposed Messina Straits suspension bridge, Italy. Reprinted from Steinman, D. B., Messina Strait suspension bridge to span 500 ft. Reprinted from *Civil Engineering*, December 1953, p. 54, with permission of ASCE.

1.10 PROPOSED FUTURE BRIDGES   37

The submerged tunnel proposal would also be a first. The idea was one of six winning projects when a competition was held in 1969 to design a fixed link across the Straits. The tunnel would "float" 131 ft (40 m) below the sea fixed to the seabed by cables.

1.10.2 Normandy Cable-Stayed Bridge

The proposed Normandy cable-stayed bridge[52,53] in France with a 2826-ft main span would carry a four-lane highway from Le Havre to Honfleur, a town on the opposite side of the Seine Estuary (Fig. 1.47). It would provide a 230-ft clearance for navigation. A preliminary design incorporates two inverted

Figure 1.47 Proposed Normandy cable-stayed bridge, France. Reprinted from Robinson, R., The French composite: A bridge for Normandy. Reprinted from *Civil Engineering*, February 1993, p. 57, with permission of ASCE.

Y-shaped, prestressed concrete pylons, one on the southern bank and the other at the edge of the navigation channel in the estuary. The design also recommends a trapezoidal steel box for the main span. A 72-ft top width would allow two lines in each direction. The projected section is 29.5 ft wide at the bottom and 10 ft deep. The designers envision a total of 200 axial cable stays fanning out from the top of the towers to anchorages in the edges of the boxes.

1.10.3 Akaghi Suspension Bridge, Japan

One of the longest proposed spans in the world is the Akaghi suspension bridge in Japan.[54] It will have a main span of 6528 ft (1990 m) and each side span will be 3148.8 ft (960 m), with the height of towers 974 ft (297 m) (Figs. 1.48 and 1.49). There will be four main cables, each $2\frac{3}{4}$ ft (840 mm) in diameter and arranged as two pairs of double cables. The current design is based on a 232,000-ksi (1600 N/mm^2) breaking strength of high-strength steel with a safety factor of 2.5. The engineers still have to decide whether to use the traditional aerial spanning method to build up the cables wire by wire in the air, or to erect prefabricated 127-wire strands.

Regarding the deck, truss-type sections were aerodynamically acceptable at 328 ft/s (100 m/s) while the boxes began to oscillate at 197 ft/5 (60 m/s). Another problem is buffeting, random oscillation due to small changes in the wind. At long spans the stiffness is less and buffeting can start at lower speeds, around 98 ft/s (30 m/s) to 131 ft/s (40 m/s) for box sections, according to tests. This in turn can lead to a further problem of fatigue.

Another worry is vortex-induced oscillation. This is extremely difficult to test in a wind tunnel since it requires a model of the whole bridge and scale factors become critical, but theory suggests that it could be a problem with the boxes at very low speeds around 23 ft/s (7 m/s) to 26 ft/s (8 m/s). As a result, it is now 96% certain that the deck will be a truss.

Despite all the high technology of the superstructure, it is actually the foundations, and particularly the caissons supporting the two towers, which have caused the greatest anxiety. Both are in water more than 115 ft (35 m) deep, one on a glacial layer of coarse gravel, the other on soft sedimentary rock. The original idea was to excavate the seabed and then sink huge steel caissons. Steel piles would then be sunk around the circumference so that the foundation could be excavated and concreted beneath the caisson. Some early model tests proved startling. They showed that under the powerful tidal currents of 13 ft/s

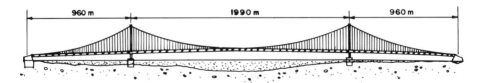

Figure 1.48 Proposed Akaghi suspension bridge, Japan.

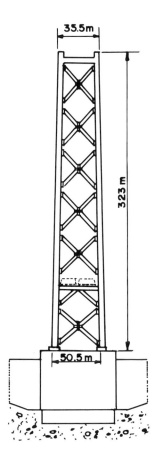

Figure 1.49 Akaghi suspension bridge—the tower.

(4 m/s) in the Akaghi Straits, scours as deep as 49 ft (15 m) could occur. The tide could easily have destroyed the bridge.

The solution was to dredge a deeper foundation. By going down 65.6 ft (20 m) below the seabed, tests showed that the total velocity at the bottom was reduced by around 20%, sufficient to remove the threat of scour. And by keeping the side slopes shallow, at 1:2.5, it is not expected to fill up with material washed in by the tide. The caissons have been changed from rectangular in shape to circular, to reduce both drag and the risk of scour. One will be 262 ft (80 m) in diameter and 230 ft (70 m) high, the other 256 ft (78 m) in diameter and 187 ft (57 m) high. The bridge is designed to resist a Richter 8 earthquake 150 km offshore, producing an acceleration of 5.9 ft/s^2 (1.8 m/s^2) at the bearing layer. Opening of the bridge is scheduled for 1998.

1.10.4 Bering Strait Bridge Project

In 1958 Professor T. Y. Lin suggested the construction of a bridge across the Bering Strait[55] to foster commerce and understanding between the people of the United States and the Soviet Union. Ten years later, Professor Lin organized the Inter-Continental Peace Bridge Inc., which was approved by the U.S. Treasury Department as a charitable corporation. The purpose of this organization is to join the shores of Alaska and Siberia.

The tides and currents in this area are not severe. The one major problem is resistance of the bridge piers against ice floes up to 6 ft (1.8 m) thick that are in constant motion during certain seasons, producing horizontal forces on the order of 5000 tons (4,536 t) or more on a pier. But there are no icebergs through the Bering Strait.

The prefabrication and construction of this bridge will be almost entirely accomplished using precast and prestressed concrete. The bridge piers could be prefabricated as one piece, then floated and sunk into position. Towers can subsequently be placed on top of the sunken piers. To reduce horizontal ice pressure on the piers, their curving slope near the water surface will help to break ice floes when they push forward and upward along the curves. Since ice pressure may develop on the piers in any direction, a circular design seems logical.

Studies indicate that the typical span length should optimally be 1200 ft (366 m). For navigational purposes, an 1800-ft (549 m) main span with a vertical clearance of 200 ft (61 m) should be provided in each channel, both east and west of the Diomedes Islands. The bridge would consist of 220 spans, 1200 ft (386 m) in length, most of which will need a vertical clearance of only 80 ft (24 m).

The substructure will be composed of 220 precast, prestressed gravity piers, each made with a shallow base raft supporting a double-curved cylindrical tower. Pier sizes may vary depending on water depth and environmental exposure. Each pier will be prefabricated in one or two pieces—the top bottle with a maximum depth of 180 ft (55 m) under water and 80–200 ft (24–61 m) above water; the bottom slab will be connected to the bottle. Each pier will be floated into position and sunk to its prepared foundation.

The optimum span length for this crossing is about 1200 ft (366 m), making the superstructure cost almost equal to that of the substructure. A single-stay cable scheme has been determined to be the most economical because the large box section can span 400 ft (122 m) between supports with a minimum number of cables. To reduce maintenance costs due to severe weather conditions at the bridge site the 1800-ft (549 m) main spans will require two cables for each cantilever.

Each 1200-ft (366 m) long deck section, with a 600-ft (183 m) cantilever on each side of the pier, can be made using 60 precast segments 20 ft (6 m) in length. These segments would be match cast in a factory and transported to a nearby assembly plant where they are joined and posttensioned together to form a double cantilever. These units would then be moved into a catamaran barge, which will transport them to the bridge site, and would be erected on top of the pier.

A maximum grade of 1% on the bridge is assumed in order to accommodate

Figure 1.50 Proposed Bering Strait bridge.

railway traffic on the bridge. The bridge section is designed to accommodate maximum anticipated traffic for 100 years or more. A box section was chosen because of its strength, efficiency, and ease of maintenance. The top deck of the box section will carry two lanes of highway traffic and will be used only under good weather conditions. A double-track railroad will be housed within the main tube, providing piggy-back transportation under all weather conditions. The lower level will house pipelines for gas transportation and other uses at the northern regions of the one for future development (Fig. 1.50).

1.10.5 Proposed Gibraltar Strait Bridge

There are several schemes for connecting Africa and Europe across the Gibralter Strait:

1. T. Y. Lin, International, San Francisco, proposed the 9.3-mi bridge across the narrowest, deepest part of the Strait—between Punta Oliveras, Spain, and Pointe Cires, Morocco—in the course of its evaluation of another 15-mi, 11-span bridge across shallower waters 12 mi to the west. The Lin bridge[56] would have two main spans of 16,400 ft each. It would exploit North Sea offshore-platform technology in construction of the piers, two in 1500 ft of water and a third in 500 ft.

The two 1970-ft high columns in each portal superstructure would cant out slightly from the bridge centerline. The main bridge cables would be set 660 ft transversely at the tops of the towers. To stiffen the structure against winds that have been measured up to 140 mph, Lin's scheme calls for drawing the main cables together with transversely placed posttensioned bracing cables so that at midspan they would be only 130 ft apart. Suspender cables would angle inward from the main suspension cables to carry the bridge deck.

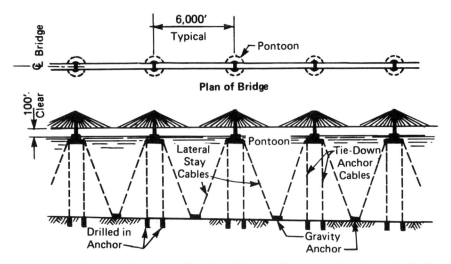

Figure 1.51 Proposed floating bridge for Gibraltar Strait crossing. Kuesel, T. R., Floating bridges for long water crossings, Annual Convention, Structural Engineers Association of Hawaii, 1984. Reprinted in Troitsky, M. S., *Cable-Stayed Bridges*, 2nd ed., Van Nostrand Reinhold, New York, p. 107.

2. A challenging innovative technical concept was proposed by Kuesel[57] for waters too deep for economical bridge piers, where multispan cable-stayed bridge piers would be built with each tower founded on a huge pontoon of 250 ft (76 m) diameter, which would be tied to the bottom by vertical anchor cables and lateral stay cables (Fig. 1.51). Kuesel proposed to build such a structure for the Gibraltar Strait, which is 15 mi (24.2 km) wide, with a meter depth to 1000 ft (305 m). A central section is proposed that will be composed of a 1000-ft (305 m) bridge span supported on tethered buoyant bridge piers. At a depth below the waves, the pier is broadened out to form a cellular pontoon structure with a net buoyancy of about 150% of the weight of the superstructure. This provides sufficient tension in the vertical anchor cables to stabilize the piers against normal tidal current, wind, and wave forces. The vertical anchor cables are deployed in a circle to give equal stability in all directions. Their tension may be monitored and adjusted through access galleries in the pontoons. A sufficient number of cables is provided to permit the removal and replacement of individual pairs of cables without endangering the stability of the system.

REFERENCES

1. Gautier, H., *Treatise on the Bridges*, André Cailleau, Paris, 1716 (in French).
2. Navier, C. L. M., *Memoir on the Suspension Bridges*, de L'Imprimerie Royale, Paris, 1823 (in French).

3. Mehrtens, G. C., *Lectures on Engineering Sciences, Steel Bridges*, Verlag von Wilhelm Engelmann, Leipzig, 1908, pp. 1–51.
4. Tyrrell, H. G., *History of Bridge Engineering*, published by the Author, Chicago, 1911.
5. Brangwyn, F. and Sparrow, W. S., *A Book of Bridges*, John Lane the Bodley Head, London, 1914.
6. Waddell, J. A. L., *Bridge Engineering*, Vol. I, Wiley, New York, 1916, pp. 1–35.
7. Watson, W. J. and Watson, S. R., *Bridge in History and Legend*, J. H. Hansen, Cleveland, 1927.
8. Block, A., *The Story of Bridges*, Whittlesey House, London, 1936.
9. Straub, H., *A History of Civil Engineering*, Leonard Hill, London, 1952.
10. Steinman, D. B. and Watson, S. R., *Bridges and Their Builders*, Dover Publications, New York, 1957.
11. Pannell, J. P. M., *An Illustrated History of Civil Engineering*, Thomes and Hudson, London, 1964, p. 209–258.
12. Naruse, Y. and Kijima, T. (Ed.), *Bridges of the World*, 2nd ed., Morikita Publishing, Tokyo, Japan, 1967.
13. Beckett, D., *Bridges*, The Hamlyn Publishing Group Limited, London, 1969.
14. Merin, O. B. (Ed.), *Bridges of the World*, C. J. Bucher Verlag, Luzern and Frankfurt/M., 1971 (in German).
15. Sealey, A., *Bridges and Aqueducts*, Hugh Evelyn Limited, London, 1976.
16. Wittfoht, H., *Building Bridges*, Beton-Verlaz Gmbtt, Düsseldorf, 1984.
17. Faustus Verantius, *Machinae Novae Fausti* Verantii, Venice, 1617.
18. Pannell, J. P. M., *An Illustrated History of Civil Engineering*, Thomes and Hudson, London, 1964, p. 312.
19. Block, A., *The Story of Bridges*, Whittlesey House, London, 1936.
20. Steinman, D. B., *Famous Bridges of the Word*, Random House, New York, 1953, p. 13.
21. Tudor, D., *Les Ponts du Bas-Danube*, Editura Academiei Republicii Romania, 1974, p. 48.
22. Block, A., *The Story of Bridges*, Whittlesey House, London, 1936, pp. 40–41.
23. Block, A., *The Story of Bridges*, Whittlesey House, London, 1936, p. 41.
24. Gies, J., *Bridges and Men*, Doubleday, Garden City, NY, 1963, pp. 13–14.
25. Gies, J., *Bridges and Men*, Doubleday, Garden City, NY, 1963, pp. 37–49.
26. Gies, J., *Bridges and Men*, Doubleday, Garden City, NY, 1963, pp. 27–32.
27. Steinman, D. R. And Watson, S. R., *Bridges and Their Builders*, G. P. Putnam's Sons, New York, 1941, pp. 85–86.
28. Steinman, D. B. and Watson, S. R., *Bridges and Their Builders*, G. P. Putnam's Sons, New York, 1941, pp. 86–89.
29. Steinman, D. B. and Watson, S. R., *Bridges and Their Builders*, G. P. Putnam's Sons, New York, 1941, pp. 81–85.
30. Gies, J., *Bridges and Men*, Doubleday, Garden City, NY, 1963, pp. 90–92.

31. Clark, E., *The Britannia and Conway Tubular Bridges*, Vol. I., Day and Son, and John Weale, 1850, London.
32. Legge, C., *A Glance at the Victoria Bridge, and the Men Who Built It*, printed and published by John Lovell, Montreal, 1860.
33. Gies, J., *Bridges and Men*, Doubleday, Garden City, NY, 1963, pp. 156–177.
34. Gies, J., *Bridges and Men*, Doubleday, Garden City, NY, 1963, pp. 216–219.
35. Steinman, D. B. and Watson, S. R., *Bridges and Their Builders*, G. P. Putnam's Sons, New York, 1941, pp. 304–307.
36. Steinman, D. B. and Watson, S. R., *Bridges and Their Builders*, G. P. Putnam's Sons, New York, 1941, pp. 282–287.
37. Block, A., *The Story of Bridges*, Whittlesey House, London, 1936, p. 104.
38. Gies, J., *Bridges and Men*, Doubleday, Garden City, NY, 1963, pp. 93–99.
39. Gies, J., *Bridges and Men*, Doubleday, Garden City, NY, 1963, pp. 183–187.
40. Trachtenberg, A., *Brooklyn Bridge*, Oxford University Press, New York, 1965.
41. Steinman, D. B. and Watson, S. R., *Bridges and Their Builders*, G. P. Putnam's Sons, New York, 1941, pp. 340–345.
42. Strauss, J. B., *The Golden Gate Bridge*, published by Golden Gate Bridge and Highway District, 1938.
43. Steinman, D. B., *Miracle Bridge at Mackinac*, Wm. B. Eerdmans, Grand Rapids, MI, 1957.
44. Talese, G., *The Bridge*, Harper & Row, New York, 1964.
45. Anonymous, New Civil Engineer reviews the history of the word's longest span suspension bridge, *New Civ. Eng. Suppl.*, May 1981, 4–22.
46. Sanson, R., Saint-Nazaire-Saint Brevin Bridge over the Loire Estuary (France), *Acier-Stahl-Steel*, No. 5, 116–167 (1976).
47. Anonymous, *The Bridge Spanning Lake Maracaibo in Venezuela*, Bauverlag GmbH, Wiesbaden-Berlin, 1963.
48. Taylor, P., *Hybrid design for the world's longest span cable-stayed bridge*, 12th/ABSE Congress, Vancouver, B. C., September 3–7, 1984, Final Report, pp. 319–324.
49. Silverberg, R., *Bridges*, Macrae Smith, Philadelphia, 1966, p. 107.
50. Overman, M., *Roads, Bridges, and Tunnels*, Doubleday, Garden City, NY, 1968, p. 107.
51. Steinman, D. B., Messina Strait suspension bridge to span 5,000 ft, *Civ. Eng.*, December 1953, 54–57.
52. Anonymous, Super starts soon, *ENR*, 18 (April 6, 1989).
53. Robinson, R. The French composite: A bridge for Normandy, *Civ. Eng.*, 56–59 (February 1993).
54. Anonymous, Longest span in the world, *New Civ. Eng.*, 18–20 (4 August 1988).
55. Lin, T. Y., Inter-continental peace bridge, *T. Y. Lin Int. Bull.* **15**(2), 4. (December 1986).
56. Anonymous, Gibraltar crossing schemes still alive, *ENR*, 22 (October 1, 1984).
57. Kuesel, T. R. *Floating bridges for long water crossings*, Structural Engineers Association of Hawaii, 1984 Annual Convention, pp. 1–13.

CHAPTER 2

BRIDGE LOCATION

2.1 INTRODUCTION

The location and layout of a bridge depends upon traffic conditions. Generally, a bridge should be located to serve the traffic best, unless other conditions are controlling. The basic principle is "bridge for the highway but not highway for the bridge." Therefore, different traffic conditions affect the general location of a bridge.[1-3]

Bridges of relatively short span are located to conform to the general location of the highway. However, for long-span bridges, the most suitable bridge location should be found after a detailed study of some alternatives. Considerations relative to the general roadway location may still be regarded as controlling factors, and the eventual adjustments in the location of the bridge are made in accordance with these requirements.

2.2 BRIDGES IN URBAN REGIONS

The important considerations related to the planning and design of urban bridges are not yet properly developed. However, it may be said that city bridges influence the landscape or transform the general character of a city. For the planning and design of modern urban-type bridges, the engineer should be guided by transportation, technical, and architectural requirements.

2.2.1 Transportation Requirements

Transportation requirements include, as the main problem, an unobstructed traffic flow of vehicles and pedestrians that may cross the river in the urban

46 BRIDGE LOCATION

environment. Poorly planned locations of urban bridges create real bottlenecks with respect to their capacity to handle urban traffic.[4]

To satisfy properly the transportation requirements, it is necessary to determine the proper width of the bridge to obtain optimum traffic flow conditions, considering the complexity of the adjoining streets and the existing, as well as the expected, traffic volume. Further, it is necessary to evaluate the most convenient types of approaches to a bridge, considering unobstructed traffic flow conditions from all the directions leading to the bridge.

Apart from the transportation problem, an economic approach should be taken into consideration if the cost of the bridge between any two points must be kept to a minimum. Because the cost of an urban bridge is generally very high, a suitable location, as well as the structural concept of the bridge, are essential to maintain bridge economy.

2.2.2 Technical Requirements

Technical requirements include:[5]

1. Determining the geometry of the proposed structure—a vertical and horizontal alignment to fit the surrounding approaches properly.
2. Choosing the main bridge system and the relative position of the deck.
3. Determining the optimum length of the spans to satisfy the hydraulic, architectural, and construction cost conditions.
4. Choosing the main elements of the superstructure and substructure, particularly the types of piers and abutments.
5. Paying attention to the details of the superstructure: railings, parapets, lighting standards, and type of pavement.
6. Choosing the most convenient materials for the bridge structure and details for aesthetic and structural reasons.

2.2.3 Aesthetic Considerations

At present, modern urban bridges are designed not only on the basis of their transportation requirements, economics, and structural characteristics, but they should satisfy the high standards of artistic appearance. The aesthetic aspect of an urban bridge is a very important factor in the overall planning of a river crossing in a city.

2.3 THE OPTIMUM LOCATION OF A BRIDGE CONSIDERING TRAFFIC FLOW

Determination of the optimum location of a bridge between cities is based on the traffic flow.[6] The origin and destination points and the volume of traffic flow are assumed to be known, where ADT = average daily traffic flow and

2.3 THE OPTIMUM LOCATION OF A BRIDGE CONSIDERING TRAFFIC FLOW

volume is equal to

$$\frac{\text{total traffic per year}}{365 \text{ days}}$$

The optimum location of crossing the river by bridge is found by the method of successive approximation from the condition of minimum traffic work. Let n points be given on both sides of the river as points of origin and destination of the traffic (Fig. 2.1). The volume of traffic flow starting from each of these points is also given. To obtain a minimum of traffic work during a trip from one side of the river to the other, it is necessary to find a certain adequate point B (bridge) on the axis of the river.

The traffic work W may be defined by the following formula:

$$W = Vl \tag{2.1}$$

where V is the volume of the traffic and l is the length of the route. Point B is the assumed location of the bridge and we trace the line connecting it with the points marked on both sides of the river (Fig. 2.2). We may denote l_{iB} as the distance between the ith point and the bridge and V_{Bi} as the corresponding traffic volume.

Similarly, in the inverse direction l_{Bi} is the distance between the bridge and the ith point and V_{Bi} is the corresponding traffic volume. The distances l_{iB} and l_{Bi} are of course, equal, or

$$l_{iB} = l_{Bi}$$

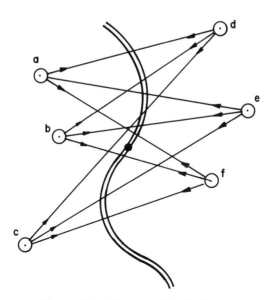

Figure 2.1 Diagram of traffic flow.

48 BRIDGE LOCATION

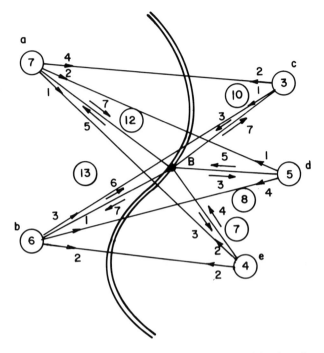

Figure 2.2 Diagram of traffic flow assuming bridge location.

The traffic work necessary to carry all traffic volumes to the bridge is

$$W_{Bi} = \sum_{i=a}^{i=f} V_{iB} l_{iB} \tag{2.2}$$

The traffic work from the bridge to the points from a to f is

$$W_{Bi} = \sum_{i=a}^{i=f} V_{Bi} l_{Bi} \tag{2.3}$$

The total traffic work is

$$W = \sum_{i=a}^{i=f} V_{Bi} l_{Bi} + \sum_{i=a}^{i=f} V_{Bi} l_{Bi} \tag{2.4}$$

But because $l_{iB} = l_{Bi}$,

$$W = \sum_{i=a}^{i=f} (V_{iB} + l_{Bi}) l_{iB} \tag{2.5}$$

2.3 THE OPTIMUM LOCATION OF A BRIDGE CONSIDERING TRAFFIC FLOW

In order to determine the length of the routes, l, we use the rectangular coordinate system, so that the Y axis is tangential to the river axis at the point of assumed crossing of the river. The X coordinate of the bridge location in the assumed coordination system is equal to zero, or $X_B = 0$ (Fig. 2.3).

Using the coordinate system shown in Figure 2.3, (2.5) takes the form

$$W = \sum_{i=a}^{i=f} (V_{iB} + l_{Bi}) \sqrt{(Y_1 - YB)^2 + X_i^2} \quad (2.6)$$

To determine the minimum value of W, we differentiate expression (2.6) by Y_B (arbitrary value), and obtain

$$\frac{dW}{dy_B} \sum_{i=a}^{i=f} (V_{iB} + l_{Bi}) \frac{Y_1 - YB}{\sqrt{(Y_1 - YB)^2 + X_i^2}} \quad (2.7)$$

For practical application, we may simplify (2.7), introducing angle α, as follows:

$$\frac{Y_1 - YB}{\sqrt{(Y_1 - YB)^2 + X_i^2}} = \cos \alpha$$

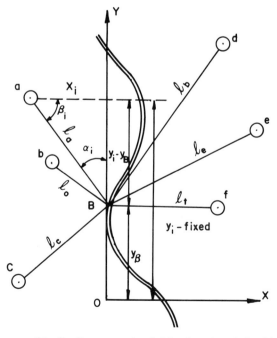

Figure 2.3 Diagram of traffic flow assuming bridge location defined by the proposed coordinate system.

Therefore, we may write

$$W = \sum_{i=a}^{i=f} (V_{iB} + l_{Bi}) \cos \alpha_i = 0 \qquad (2.8)$$

or

$$W = \sum_{i=a}^{i=f} V \cos \alpha_i = 0 \qquad (2.9)$$

Equation (2.9) indicates the condition for the traffic work W to be minimum. By solving this equation using the method of successive approximation, we may find the optimal location of the bridge, considering traffic flow.

2.4 NUMERICAL EXAMPLE

Determine the optimum location of a bridge from the condition of minimum traffic work. The following data and average yearly traffic flow per day (ADT) are shown in Figure 2.2.

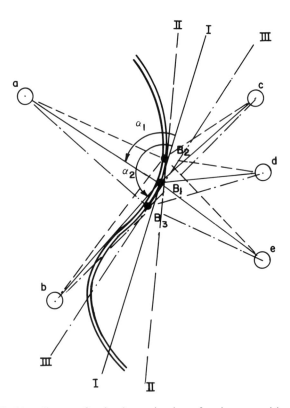

Figure 2.4 Working diagram for the determination of optimum position of the bridge.

2.4 NUMERICAL EXAMPLE

TABLE 2.1 Location B_1

i	ΣV	α_i	$\cos \alpha_i$	$\Sigma V \cos \alpha_i$ +	$\Sigma V \cos \alpha_i$ −
a	12	75°	$\cos 75° = \sin 15° = +0.259$	3.10	
b	13	155°	$\cos 155° = \cos(180° - 25°) = \cos 25° = -0.906$		11.80
c	10	327°	$\cos 327° = \cos(360° - 33°) = +\cos 33°$	8.39	
			$= +0.839$		
d	8	288°	$\cos(270° + 18°) = \sin 18° = +0.309$	2.47	
e	7	250°	$\cos(270° - 20°) = -\sin 20° = -0.342$		2.40
			Summary	13.96	14.20
				−0.24	

In the following we assume the arbitrary positions of the bridge crossing, namely, B_1, B_2, B_3, and so on. For each bridge position we introduce the rectangular coordinate system so that the Y axis is tangential to the river axis at the point of assumed crossing of the river. Further, considering the first crossing B_1, we evaluate the angles α_i and values $\cos \alpha_i$. If

$$\sum_{i=a}^{i=f} V \cos \alpha_i = 0$$

this indicates the ideal position of bridge crossing. Figure 2.4 shows the working diagram. Tables 2.1–2.3 show all values for assumed bridge locations B_1, B_2, B_3.

TABLE 2.2 Location B_2

i	ΣV	σ_i	$\cos \sigma_i$	$\Sigma V \cos \alpha_i$ +	$\Sigma V \cos \alpha_i$ −
a	12	73°	$\cos 73° = \sin 17° = +0.292$	3.50	
b	13	150°	$\cos(180° - 30°) = -\cos 30° = -0.866$		11.26
c	10	310°	$\cos(360° - 50°) = +\cos 50° = +0.643$	6.49	
d	8	270°	$\cos 270° = 0$	0	0
e	7	230°	$\cos(270° - 40°) = -\sin 40° = -0.643$		4.50
			Summary	9.99	15.76
				−5.77	

TABLE 2.3 Location B_3

i	ΣV	α_i	$\cos \alpha_i$	$\Sigma V \cos \alpha_i$ +	$\Sigma V \cos \alpha_i$ −
a	12	100°	$\cos (90° + 10°) = -\sin 10° = -0.174$		2.09
b	13	168°	$\cos (180° - 12°) = -\cos 12° = -0.978$		12.71
c	10	350°	$\cos (360° - 10°) = \cos 10° = +0.985$	9.85	
d	8	320°	$\cos (360° + 40°) = \cos 40° = +0.766$	6.13	
e	7	278°	$\cos (270° + 8°) = -\sin 8° = -0.139$	0.97	
			Summary	16.95	14.80
				+2.15	

REFERENCES

1. Waddell, J. A. L., *Bridge Engineering*, vol. II, Wiley, New York, 1916, pp. 1088–1092.
2. Johnson, L. C., How to determine bridge location, *Heavy Construction News*, 3, 4, 36, (October 20, 1960).
3. Campbell, M. E., Traffic factor in bridge planning, *Traffic Q.*, 132–162, (1969).
4. McCullough, C. B., Highway bridge location, United States Department of Agriculture Bulletin No. 1486, United States Government Printing Office, Washington, DC, 1927, pp. 2–12, 27–32.
5. Doten, H. L., Principles and techniques of highway bridge surveys, Maine Technology Experiment Station, Bulletin No. 41, University of Maine, Orono, ME, June 1946.
6. Martusewicz, J., The optimum location of a bridge from the point of view of the traffic engineering, *Arch. Inzynierii Ladowej*, **IX** (Z.1), 117–125, (1963) (in Polish).

CHAPTER 3

BRIDGE LAYOUT

3.1 SQUARE OR SKEW BRIDGE

When the general location of a bridge is to be decided there are a number of variables to be considered. The most important is the layout of the bridge with respect to the topographic crossing.

In the early days of highway systems development, highway standards were extremely low and the cost of bridges represented a very high proportion of the total cost of the highway. Consequently, structures were nearly always constructed at an ideal location so that the span was the shortest possible, the foundation conditions the best obtainable, and the crossing a square one if at all possible. As a result, the alignment of the highway often suffered. With the advent of the car, standards changed, but problem did not.

With every new bridge project there are two somewhat conflicting views—those of the bridge designer and those of highway engineer.

1. *View of the Bridge Designer.* The square crossing of a river, canyon, or railway is nearly always preferred. When confronted with a skewed alignment, the bridge designer prefers a square bridge to a skewed bridge. These preferences are based on a natural desire to bridge a gap as economically as possible. The skewed structure is considered an "abomination" by the bridge engineering profession in the opinion of J. A. L. Waddell.[1]

2. *The Layout of Small Bridges.* It is obvious that for small structures the layout of the bridge itself is generally of secondary importance compared to the general alignment. The highway designer often locate(s) the highway so that the structures have to accommodate partially or sometimes fully curved alignment.

54 BRIDGE LAYOUT

3. *The Layout of Extraordinary Span and/or Length Bridges.* As the structure increases in length and in importance, however, the layout of the bridge becomes increasingly important.[2] Who then should decide the location of a section of highway involving a bridge? How much must the highway engineer be influenced by the wishes of a bridge designer? This question is at the heart of the problem: At what span length, or at what cost relationship between structure and highway, should the structure or highway dictate the layout? It is hard to answer this question; we can only give an indication of the range. To achieve a minimum cost of crossing the river, the cost of highway plus the cost of structure must be kept to a minimum. This involves a compromise between the bridge designer and highway engineer. This minimum cost, however, is influenced by many factors. One of the most important is the angle of crossing.

3.2 ANGLE OF CROSSING

Angle of crossing has a great effect on the length of crossing. Let us compare the cost of a square bridge with that of a bridge on a skewed alignment (Fig. 3.1) where

L = length of a square bridge
L_s = length of a bridge on a skewed alignment
ϕ = an angle of skewed alignment

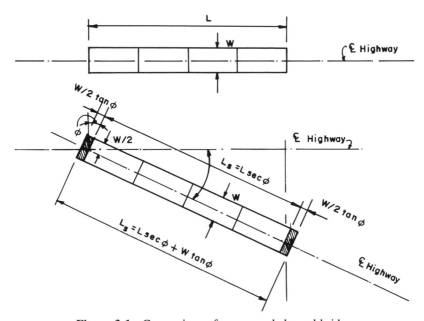

Figure 3.1 Comparison of square and skewed bridges.

3.2 ANGLE OF CROSSING

The length L_s of the structure on the skewed line is

$$L = L_s^1 \cos \theta$$

$$L_s^1 = \frac{L}{\cos \theta} = L \sec \phi \qquad (3.1)$$

Usually we find that the layout necessitates lengthening the bridge to bring the abutments back to the line of the square bridge abutments. This further increase in length amounts to $w \tan \phi$, where w is the out-to-out width of the structure.

Therefore, the total length is

$$L_s = L \sec \phi + w \tan \phi \qquad (3.2)$$

If both structures comprise a series of spans using the same span as a module and the same substructure units, the cost per foot of both can be considered equal and the cost of the structure on the skewed alignment would be $\sec \phi$ times greater than that of the square structure.

When we say

A = the cost of the square bridge per foot for superstructure in dollars
B = the cost for substructure in \$/ft

then the overall cost of the square bridge is

$$C = L(A + B) \text{ in \$} \qquad (3.3)$$

For the skewed structure that cost would be

$$C_s = (L \sec \phi + w \tan \phi)(A + B) \qquad (3.4)$$

The ratio is

$$\frac{C_s}{C} = \frac{L \sec \phi + w \tan \phi}{L} = \sec \phi + \frac{w \tan \phi}{L}$$

But the value $(w \tan \phi)/L$ is generally small, therefore

$$C_s = C \sec \phi \qquad (3.5)$$

If $\phi = 45°$, then $\sec 45° = 1.414$ and $C_s = 1.414C$ or more than 40% increase in cost.

3.2.1 Skewed Substructure

Frequently, the substructure of the bridge must be skewed and this generally results in the skewing of the whole structure (Fig. 3.2). This type of structure is often necessary when crossing streams and rivers to minimize scour, ice pressure, and the catching of drift. It is also used extensively for overhead crossing of railroads, highways, and divided highways.

Assuming again that both structures use the same span as a module, the cost of the superstructure, though basically the same, has increased because of the cost of details by about 5%, depending upon the details and skew angle.

Meanwhile, the cost of the substructure can be as much as $\sec \phi$ times greater than that of the square crossing. Using the same definition for A and B as in (3.3), the overall cost of the square bridge is also $C = L(A + B)$ in $.

For the skewed structure the cost would be

$$C_s = L_s(1.05A + B \sec \phi) \text{ in } \$ \qquad (3.6)$$

or in terms of L,

$$C_s = L \sec \phi \, (1.05A + B \sec \phi) \text{ in } \$ \qquad (3.7)$$

In the examples given above we assumed a constant span module, and therefore a constant cost per foot of superstructures. If conditions are such that the span or spans used for the skewed structure must increase, then the cost of the superstructure must change. Often structures must comprise the same number

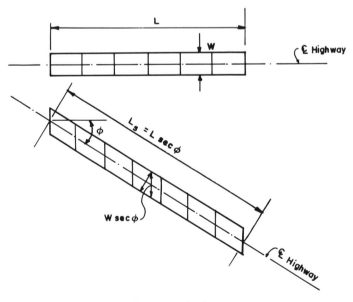

Figure 3.2 Skewed substructure.

of spans for a skewed arrangement, so that the span length must be multiplied by sec ϕ. This increase in span length increases the cost per foot of the superstructure, and also imposes a greater load on the substructure, which in turn may have to be increased in size and therefore in cost.

3.2.2 Alternative Layouts of the Bridge

The choice of a bridge layout is determined on the basis of comparative estimates of the cost of bridge and highway for the various alternatives. Frequently there is an ideal bridge site A for square crossing that is located some distance from the site B having skew crossing that has fitted into the overall highway location[3] (Fig. 3.3). The cost of the bridge at site A would be considerably less than the one at site B, perhaps due to its shorter length or cheaper (economical) foundation. However, the ideal site for the bridge is only reached at the expense of extra highway. The problem is solved by determining the difference in cost of the structures and assessing whether the added highway costs would be less than this difference.

Let us consider only two alternative highways going from M to N (Fig. 3.3). Alternative A—the square crossing comprises H_1 feet from M to N and a length of bridge L_1 feet. Alternative B—the direct highway is H_2 from M to N with a bridge of length L_2 feet. Let us assume that for any set of alternative routes the cost per foot is constant:

K_B = cost of the bridge per ft
K_H = cost of the highway per ft

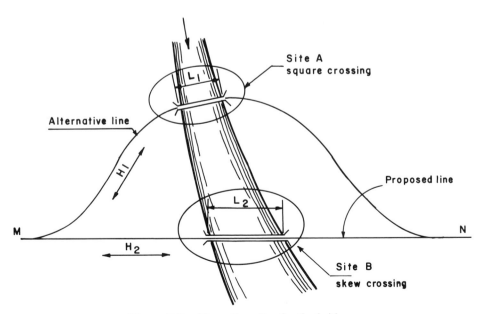

Figure 3.3 Alternative sites for the bridges.

58 BRIDGE LAYOUT

In order to justify the square crossing route (Alternative A), its cost must be equal to or less than the cost of the other route. Expressed in the form of an equation,

$$(H_1 - L_1)K_H + L_1 K_B \leq (H_2 - L_2)K_H + L_2 K_B$$

$$(H_1 - H_2)K_H \leq (L_2 - L_1)K_B - (L_2 - L_1)K_H \tag{3.8}$$

$$(H_1 - H_2)K_H \leq (K_B - K_H)(L_2 - L_1)$$

or dividing by K_H,

$$H_1 - H_2 \leq \left(\frac{K_B}{K_H} - 1\right)(L_2 - L_1)$$

Denoting the relationship between the cost per foot of bridge and the highway as

$$K = \frac{K_B}{K_H} \tag{3.9}$$

we obtain

$$H_1 - H_2 \leq (K - 1)(L_2 - L_1) \tag{3.10}$$

From (3.10) it is apparent that the squared crossing route is economically justified if the length H_1 is not greater than the length of the other route H_2 by an amount $(K - 1)$ times the difference in the span lengths $(L_2 - L_1)$. By substituting numerical values in this equation, an approximate answer can readily be obtained as to the relative economy of the square crossing route compared to the skewed crossing route.

3.2.3 Summary

Bridges on a skewed alignment are more expensive than those on a square alignment. The economic justification for a square alignment over a skewed alignment increases with the length to be spanned and the angle of skew proposed for the skewed alignment. Thus cooperation between the bridge designer and the highway engineer is very important.

Other Factors. It should be realized that there are other costs influencing this comparison, but in most cases, they are not dominant:

1. The road users' cost over the extra length of the square crossing route may bear consideration in areas of heavy traffic volume.

2. Terrain is one variable that probably plays the most important role in highway location.
3. The standards of construction is another variable in highway location.

Just as there are factors that influence the cost of highway construction, so there are many factors that influence the cost of a bridge:

1. Grade must be established to provide the necessary clearance for the depth of the structure.
2. The foundation conditions play an important part in the design and cost of the bridge as we will see in the following section.

3.3 BRIDGE APPROACHES

On small streams, the route of the road should not be changed within the limits of the bridge passage.

On the middle and large rivers the bridge approaches should deviate as little as possible from the general direction and form a continuous straight line with the axis of the bridge.

The route of the road at the bridge approach should be, as near as possible, a straight line. In the case of a change in the road direction, the largest possible bent radius should be applied, so visibility will not be limited. Bridge approaches should not have a steep gradient, which decreases road capacity. Sharp gradient bends with small radii just before the bridge are particularly dangerous for traffic.

3.4 SITE INVESTIGATION

3.4.1 General Data

After general location and layout of the bridge are established on the map in the office, the engineer may prepare the preliminary design. However, before preparation of the preliminary design, it is necessary to learn about the future bridge site, particularly the future substructure conditions, considering river characteristics.

It is often said that "the bridges are built under the water." Therefore, the first step in considering the design and construction of a bridge is a detailed site investigation. The type, span length and cost, and to some degree the appearance of the bridge are determined by the results of these investigations. The whole work consists of two parts which complement each other, namely, office work and field work.

3.4.2 Office Work

1. It is necessary to have a complete topographical map of the site indicating the location of the bridge.
2. The geometry of the river in the vicinity of the bridge must be shown to a suitable scale on the site plan.
3. The layout of the bridge should be shown on the site plan.

3.4.3 Field Work

When the approximate location is fixed, a special survey and a complete and extensive report about the site must be completed.

Site Investigation. The bridge designer should investigate the physical nature of the site, for example, the banks of the river. Possible slides must be surveyed and shown on the drawings. Photographs, particularly color slides, will give an impression of the site characteristics. Local outcrops of rock are of interest. The engineer should study the geographical situation—the availability of materials, equipment, and workshop facilities as well as local transport problems.

Foundation Conditions. Foundation conditions, either at the location of crossing or location of piers in a crossing, may be controlling factors. Unsatisfactory conditions at a projected site, due to caverns in rock, geological faults, or soft underlying material to any great depth, may, perhaps, be avoided or improved by shifting to an adjacent site. With the site fixed, borings may show that certain pier locations are better than others.

Each pier and abutment of a bridge must be considered separately. Extrapolation from borings information by drawing soil layer profiles is a risky but often necessary practice. Shallow water, good material at moderate depths, and rock outcrops, may influence pier locations and span lengths. River conditions may indicate or require a particular erection method for certain spans and thus influence choice of span type.

Deep or swift water or rock bottom with little overburden, making falsework dangerous or expensive, would indicate cantiliever erection, which is better adapted to continuous spans or cantilevers than to simple spans.

Survey Report. Printed standard forms for reports are recommended to insure coverage of every possible iterm. These reports may contain additional sketches and notes, full-scale maps, and profiles. The bridge survey report should include accurate data on the channel or waterway for all stages of water, the foundation conditions, and the stream characteristics. Data on adjacent structures on the stream, particularly their waterway opening, are especially important.

REFERENCES

1. Waddell, J. A. L., *Bridge Engineering*, vol. II, Wiley, New York, 1916, pp. 1210–1218.
2. Johnson, L. C., How to determine bridge locations, *Heavy Construction News*, 3, 4, 36 (October 20, 1960).
3. McCullough, C. B., Highway bridge location, United States Department of Agriculture Bulletin No. 1486, United States Government Printing Office, Washington, DC, 1927, pp. 20–30.

CHAPTER 4

CROSSING THE RIVER

4.1 LAYING OUT THE BRIDGE SPANS

During the development of the bridge crossing project it is necessary to solve a number of problems. Among them are choice of the bridge location and its elevation, determination of the design flow discharge in the river, laying out of the bridge spans, choice of the spans system, and determination of the foundation types and their depth. Solution of the problem of the bridge layout in the flood plain depends on the character of the river to be crossed.

When crossing a relatively small river, the location of the bridge follows the direction of the highway. For larger rivers the location of the bridge is usually at a right angle to and at a narrow section of the river or at the section of minimum floodplain. In this case the river channel should be straight and not changed; geological conditions should be convenient for the foundations of the piers.

To satisfy the abovementioned conditions it is possible to deviate the direction of the highway from its basic alignment. The final solution may be accepted after analysis of the different alternatives of the bridge crossing.

The determination of the design flow discharge is estimated after observing discharges for a number of years. Statistical results of such observations, based on the theory of probability, permit a determination of the value of average discharge. For major highway bridges on primary highway road classifications, the design discharge for design is generally that for a 50 year return period; i.e., a 2% chance of occurrence in any given year. Often the 100 year return period potential damage to surrounding land areas is examined and established as the design flow discharge. For important bridge locations the design flow discharge may exceed the 100 year return period. Considering the established design flow discharge the designer determines the necessary working area of the stream. The working area depends, among other factors, on the calculated discharge, average velocity of the stream at the opening, and value of the contraction of the stream. The economical solution is determined by comparing alternative bridge crossings. The cost of each crossing according to each alternative should be the total cost of all bridge elements and bridge approaches. For important river crossings the average stream velocity may be insufficient to establish the working area of the stream. In such instances a more rigorous analysis is needed including finite element flow and back water analyses.

For each alternative bridge crossing, the most economical and technically advantageous bridge scheme should be found. This scheme is found by developing a few alternatives and determining the length of the spans, the system of the spans and their general dimensions, and the type of piers and their foundations. During development of the bridge alternatives conditions of fabrication and erection of the spans, methods of pier construction, time of construction, and architectural merits of the structure should be considered.

To satisfy all the conditions noted above, it is necessary to compare a few bridge alternatives. This comparison utilizes experience in the design of similar structures, data from engineering science and practice, and existing solutions.

The choice of bridge scheme is a creative problem, connected to independent work, which cannot be based on recipes. A textbook may only show the steps to the solution of separate problems and state general considerations regarding the essence of a solution. To compose bridge alternatives it is necessary to first establish the elevation of the deck. On rivers open to navigation this elevation depends on the lateral and vertical clearance requirements of the navigation channel.

Along a multispan bridge there should be one major long span, to minimize the potential for ship impact which could result in collapse or damage to the bridge that would necessitate closing the bridge and/or navigation channel for repairs. Such an action would impose an economic hardship to shipping or vehicular traffic on the bridge. Further, damage to a ship or bridge may result in hazardous spills and the ensuing cost of cleanup. In the case where hazardous material has settled to the bottom of the river or bay, longer spans may be more economical than environmental containment of excavated material for foundations and disposal of contaminated material.

4.2 PRINCIPLES OF BRIDGE CROSSING DESIGN AND ITS STRUCTURES

4.1.1 General Data on Water Passage and Bridge Crossing

All territory from which water flows into a river is called a basin. In a basin, a river valley is isolated in the shape of terraces created after the changeable processes of sedimentation and washout. The river flows along the narrow part

64 CROSSING THE RIVER

of a valley or river channel. The low part of the valley and the shores adjoining the valley which are periodically flooded represents the floodplain.[1]

The cross-sectional area of the river, perpendicular to its flow, is called the free-flow section. The quantity of water flowing through this section in units of time is discharged through the water passage.

For the constantly acting water passages we consider: the level of high water (HW) during the spring flood, calculated level of high water (CHW), calculated shipping level (SLW), and low level of water during the summer period (LW).

At a crossing of a water passage by highway or railway bridges, a complex of engineering structures may be constructed and thus constitute the bridge crossing. This bridge crossing may consist of the following elements (Fig. 4.1):

1. The *bridge*, which crosses over the high-water part of the free-flow section, and which may pass over part of the channel and flood plain as well,
2. *Approaches* to the bridge, consisting of embankments or an approach viaduct structure, and
3. *Regulation and reinforcing structures* to improve conditions of river flow.

To choose the best location for the bridge crossing, it is necessary to consider the whole complex of the river characteristics that influence the cost of

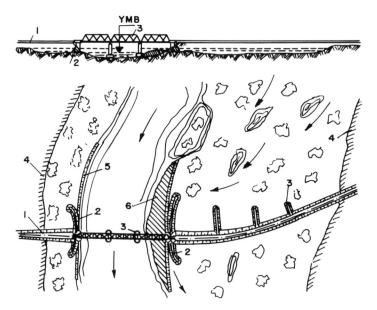

Figure 4.1 Cross-section and plan of bridge crossing: (1) approach embankment; (2) current directing dike; (3) bridge; (4) limit of flood plain; (5) reinforcing of the shore; (6) cross arm.

the structure. Such characteristics include geological conditions, which define type and depth of foundation for bridge supports; topographic conditions, which determine work volume for building approaches to the bridge; hydrological conditions, particularly high-water flood width and width of the channel velocity of the water current, which define length of the bridge, and work elements for river regulation protection of the flood plain.[2] A bridge crossing, which unduly constrains water flow may result in scour of the river bottom and lead to failure of the structure.

During a comparison of possible alternative arrangements for the bridge crossing, it is necessary to find the optimum degree of river constraint due to the crossing. The most expedient means is to total expenses for construction and use of the structure. A bridge crossing should satisfy definite requirements for shipping under the bridge. Before designing a bridge crossing, preliminary field investigations are necessary to collect the abovementioned materials and data on the river as well as topographical, ground, and geological conditions at all possible alternatives of bridge crossing. For small and middle span bridges, location of the crossing is defined by the general direction of the highway, but for larger bridges, as a rule, a few possible alternatives of bridge crossing are planned. In many densely populated areas or locations otherwise constrained by existing physical features the crossing may be defined by such features.

4.2.2 Exploration of Bridge Crossing

During the preparation period existing materials, which characterize the location of the bridge crossing and regime of the river under consideration, are collected and studied. On the basis of maps, the location of the crossing is chosen and the surveying party and exploration are organized.

To learn about river characteristics and other conditions of crossing location, literature and reference materials are studied, such as data from projects and experience from the use of other existing bridge crossings on the river under consideration. During field investigations, a survey is done to provide more detailed data for a relief of the river valley, and a better choice of proposed crossing location. General and detailed plans are surveyed and cover the whole area of possible alternatives for the river crossing. Detailed plans are necessary for the design of regulation structures, for reinforcing of the shores, and for temporary structures. As a rule, for each basin the surface flow based on runoff from a design rain intensity and/or spring ice thaw waters is defined.

To determine the opening for middle and large bridges,[3] it is necessary to know the velocity of the river current during the inundation period. In the channel of the river it is greater, and over the flood plain and at the shores it is smaller. The longitudinal slope of the river and the roughness of the channel bottom influence the velocity of the flow at the bridge crossing. Discharge of the water is changed depending on climatic conditions and on the largest calculated discharge in 50 or 100 years. The design of bridge openings, flood plain embankments, and regulation structures under the action of water dis-

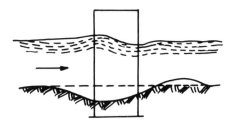

Figure 4.2 Scheme of local scour at support.

charge is carried out using calculated discharges and corresponding water levels. Construction of a bridge crossing causes constraint of the free-flow section and a substantial change in its regime, as well as deformation of the channel.

Pier supports of rectangular shape, due to their great resistance to the water flow, cause great washout of the river bottom in comparison with the supports having rounded and pointed front and rear sides. Therefore, the region of water flow should be considered when choosing the shape of supports. Local scour is probable close to supports (Fig. 4.2).

The constant pressure of the stream causes afflux of the water, or increase of the level in front of the bridge (Fig. 4.3). With greater constraint of the river cross-section and smaller scope of the water surface we have the distribution of afflux over a greater distance. Afflux must be considered when designating the embankment height and regulation dikes.

During the process of exploration, an important component is the geological investigation to determine the physical–mechanical properties of the ground as a basis for bridge supports. Such investigations are performed by borings.

4.2.3 Choice of Location for Bridge Crossing and Design of Bridge Opening

Following the results of exploration, the location of the bridge crossing is chosen and its opening is designed.[4,5] For large bridges, a few alternatives for

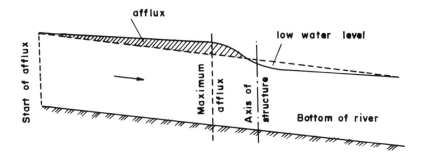

Figure 4.3 Longitudinal profile of water surface during creation of the affluent in front of the bridge.

possible crossing location are usually considered. When choosing the optimum solution the following requirements are considered:

1. The proposed route of bridge crossing should be as close as possible to the general direction of the designed route.
2. Considering engineering and geological conditions, the section of the river at the crossing location should be most convenient for construction of supports and approach embankments to the bridge.
3. When choosing crossing location it is necessary to avoid or at least design for locations that may be undergoing change.
4. The crossing should be at the most narrow section of the river and the smallest width of flood plain, if possible.
5. The crossing should avoid the island and branch arm of the river because they may change the flow of the river.
6. The axis of the bridge should be normal to the river flow in the channel and flood plain.
7. On navigation rivers the bridge should be located on the rectilinear section of the channel.
8. Because bridge crossings over large or medium rivers is expensive, the design of highway direction should be in accordance with the crossing location where possible; however, in case of railways the location of bridge crossing should coincide with the general direction of the railway line.
9. The cost of the bridge depends on the materials, length, number, and type of supports. The length of the bridge and number of supports are defined by the bridge opening. During design of the bridge opening the following basic data are considered:
   a. Provision for traffic safety on the bridge, which is achieved by creating safe conditions for bridge crossing and stable conditions for the passing of high water.
   b. Do not permit great affluent of the water created by the narrow opening.
   c. Maintain unobstructed shipping on the river.
   d. For the stability of the structure, the opening of the bridge should not be less than the width of the navigation channel.

4.2.4 Sequence of Bridge Design and Comparison of Alternatives

After determination of bridge opening its scheme is designed considering the approaches to the bridge. The following should be considered: (a) conditions of navigation, (b) cost of bridge construction, and (c) expenditure for materials and labor.

For navigation rivers, the scheme of the bridge depends on the clearances under the bridge. Multispan bridges should have at least one long navigation

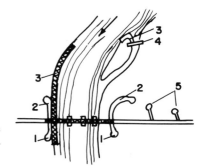

Figure 4.4 Regulation structures: (1) spur dike downstream; (2) upstream; (3) shore-reinforcing; (4) dam; (5) cross arm.

span. There are often multiple accommodations—for example navigation versus recreational type opening requirements.

For small rivers, where the size of the shipping opening is small, often it is economically expedient to have equal spans along the whole length of the bridge.

A technical-economical comparison of alternatives for a large bridge crossing is very complicated. Basic factors for comparison as well as the cost of building and the length of the works for construction of the bridge crossing are utilized. The expedience of the bridge project is determined by comparing the corresponding technical and economical indices.

After comparing the bridge crossing alternatives by considering the building and architectural data, the most rational (optimal) alternative is chosen.

4.2.5 Bridge Approaches—Regulation and Bank Protection Structures

The elements of a bridge crossing include the approaches to the bridges located on or over the flood plain, as well as regulation and bank protection structures which direct the passing of the high water under the bridge.[6]

River streams having free flow over the flood plain at the approach embankment change direction and flow toward the opening of the bridge. This may cause scour, which endangers the stability of bridge crossing elements. To provide safe flow of the water under the bridge regulation structures are built such as spur dikes (Fig. 4.4). These dikes allow smooth insertion of the overbank flow into the river flow. They may be curvilinear or rectilinear (Fig.

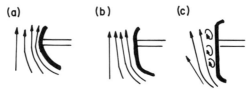

Figure 4.5 Schemes of spur dikes: (*a*) curvilinear; (*b*) rectilinear with straight insertion; (*c*) rectilinear.

4.5). Curvilinear spur dikes are built on the rectangular section of the river, with straight insertion at the curvilinear configuration of the channel.

Spur dikes are embankments from sandy ground, having trapezoidal configuration. The side slope from the river side is 1:2 and from the flood plain no more than 1:1. Slopes from the river side of the spur dike are reinforced by concrete slabs on stone paving.

4.3 GENERAL CONDITIONS FOR THE LAYOUT OF A BRIDGE ACROSS A RIVER

A correctly designed bridge passage over the river should satisfy the following hydraulic conditions:

1. Secure the passing of the discharge of water and ice through the opening with the least disturbance of the regime of the stream.
2. Let ships and barges pass conveniently and safely under the bridge on navigable rivers.

These requirements can be fulfilled only when the location of a stream crossing is correctly selected—usually a difficult task under various local conditions.[7,8]

The general conditions that determine the layout of a bridge crossing the river may be summarized as follows:

1. *Crossing of the main channel and the valley flats.* The valley flats should be crossed at their narrowest point and the main channel at its widest (Fig. 4.6).

2. *Direction of the river crossing with respect to the main channel and valley flats.* The selected sector of the stream should facilitate the designing of a route perpendicular both to the main channel and the valley of the stream. The span of the bridge should be so located as to allow the main stream to pass through the center opening of the main span (Fig. 4.7). In case we locate the bridge at Section I-I, the high-water direction will be skewed with respect to

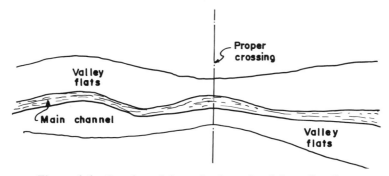

Figure 4.6 Crossing of the main channel and the valley flats.

70 CROSSING THE RIVER

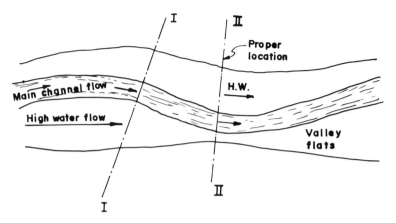

Figure 4.7 River crossing with respect to main channel and valley flats.

the pier and this may cause scour or erosion. Therefore, it is better to locate the bridge along the axis II–II, approximately perpendicular to the flow of the high water in the valley flats.

3. *Main river and tributary.* In view of the possibility of sediment accumulation, a bridge should not be located directly below the mouth of a tributary (Section I–I, Fig. 4.8). Also, the bridge should not be located close upstream on the main river flow (Section II–II, Fig. 4.8). Therefore, a stream sector should be selected where there are no side branches, so that only one bridge need be built.

4. *Permanency of channel.* Shifting or meandering streams frequently necessitate a longer crossing than would be necessary to carry maximum flood

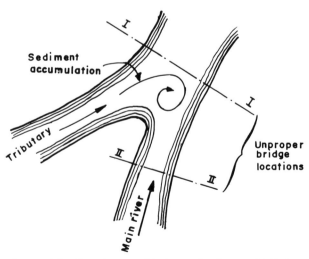

Figure 4.8 Crossing considering main river and tributary.

4.3 GENERAL CONDITIONS FOR THE LAYOUT OF A BRIDGE ACROSS A RIVER 71

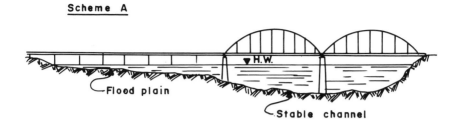

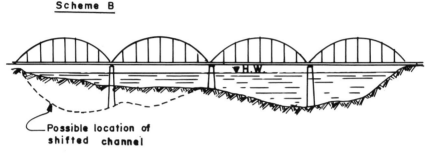

Figure 4.9 Crossing of the river. Alternative schemes considering permanency of channel.

discharge. Crossings over such streams are usually expensive, due to the long span and deep pier construction from side to side of the flood plain.

Figure 4.9 shows some typical schemes (A and B) considering different river and channel conditions. Scheme A is a typical crossing of a main channel having a stable right-hand bank and its entire flood plain on the other side.

If the main channel is stable and permanent, two river spans will be used and the balance of the flood plain may be bridged with viaduct construction at considerably less cost per linear foot.

If the stream channel is likely to shift from side to side of the flood plain during the estimated life of the bridge, it is necessary to construct Scheme B. In this case, three stream piers and four trusses are required compared to two piers and two trusses in Scheme A.

5. Large skew bridges can be built only in exceptional cases, since the piers are exposed to the impact of ice, which may cause the formation of ice barriers (jams).

6. The region of the stream sector selected should not be subject to great changes during a year. The region is the most homogeneous in straight and deep sectors.

7. On a river in its natural state, there is no place fulfilling the requirements indicated above. The necessary conditions may be achieved by artificial means such as building realignment structures, reinforcing river banks, and so on.

8. *Correction of the stream.* For crossing of small rivers and streams it is advisable not to curve the route, particularly on the more important roads.[9] In

72 CROSSING THE RIVER

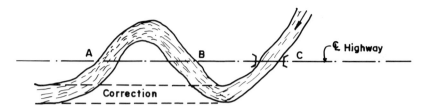

Figure 4.10 Correction of the river channel.

such cases, the direction of a stream should be changed, as shown in the following examples.

    a. Correction to reduce the number of bridges (Fig. 4.10).

    b. Correction by building an artificial river channel at right angles to the axis of a road route (Fig. 4.11). This requires careful consideration of the hydraulics of the "corrected" channel.

9. *Stream maintenance.* The channel of each stream or other natural waterway should be kept clear so that the water will be allowed to flow normally. Measures should be taken to keep logs, trees, brush, or trash from becoming lodged against bridge structures. Such debris, if unattended, can alter the course of the stream and also cause scouring and undermining.

Stream banks in the bridge vicinity should be checked for erosion. In order to minimize erosion, widening the channel may be considered to permit discharge of the same volume of water with a lower velocity.

Accumulation of silt would reduce the free board between the water in a stream and the bottom part of the piers. Deposited material should be removed from the stream bed before serious damage occurs.

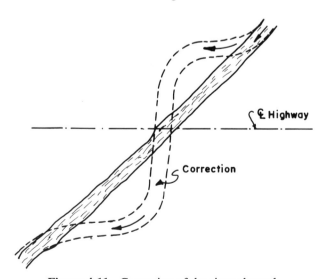

Figure 4.11 Correction of the river channel.

During a heavy rainfall, a stream may carry trees, debris, bushes, and trash which hit with great force against a bridge. Accumulation of debris on the upstream side of the bridge should be cleared and dislodged to prevent flooding.

10. *Degradation and aggradation.* Degradation is the gradual lowering of the stream bed material in the channel. Degradation may occur naturally, caused by erosion or by the following factors: stream straightening, stream flow, and reduction of sediment loads. Sometimes, owing to considerable stream bed lowering, the pier foundations should be checked for lateral slope instability. Aggradation is the deposition of sediment into the river channel. A rise in the stream bed causes the water surface level to raise. This may produce reduction of the waterway opening and lead to possible scour.

4.4 COMPUTING OPENINGS OF BRIDGES

4.4.1 Definitions

The opening of a bridge L_0 is the distance between the front walls of abutments measured on the water level acceptable for determining the size of the opening. The same distance, less the total width of all the bridge piers, is called the clear span of a bridge[10] (Fig. 4.12). The opening of the bridge is

$$L_0 = b_1 + b_2 + b_3 + \cdots + a_1 + a_2 + \cdots \quad (4.1)$$

The clear span is

$$L_c = L_0 - a_1 - a_2 = b_1 + b_2 + \cdots = \sum_{i=i}^{i=n} b_1 \quad (4.2)$$

where

b_1, b_2 = the individual clear spans
a_1, a_2 = width of the piers

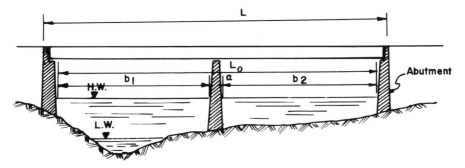

Figure 4.12 Opening and clear span of the bridge.

74 CROSSING THE RIVER

The length of a bridge L is the distance between the back walls of the abutments or the length of the superstructure of the bridge.

4.4.2 Computation of Bridge Opening

Computing the bridge operation is accomplished by:

1. Determining the smallest cross-section under the bridge necessary to pass the water with an adopted velocity. The area of the cross-section filled with flowing water is called the discharge section area or the active cross-section.
2. Finding the appropriate value of the size of the opening that correlates to the existing cross-section of the channel within the limits of a bridge passageway.

To determine the bridge opening we consider the following two cases:

1. *The determination of the bridge opening on the assumption that the discharge section area will not contract.* This condition holds if the bridge opening is equal to or larger than the width of a valley flat valid for computing the discharge. No headwater will occur in this case. Therefore, no directive dikes are built there, nor is any river correction work performed. Although the cost of building a bridge longer than the width of a valley flat is usually high, such a solution is at times advisable and even necessary. Mountain rivers having high current velocities and variable shifting channels are often crossed without contraction of the cross-section. If the ground in valley flats is unsuitable as a support for the bases of an embankment—for instance, if it consists of marshy peats—a bridge instead of embankments should be built in valley flats, because to stabilize the embankment would be expensive.

If it is necessary to build a river crossing in a hurry (war, emergency), then, too, bridges are built instead of high embankments. They can be constructed much more quickly than high embankments, especially when timber is used for construction.

2. *The determination of the bridge opening on the assumption that the discharge section area will contract.* Usually, it is less costly to build an embankment than a bridge. For this reason, bridges shorter than the width of valley flats are usually designed and earth embankments are built to both ends. In such cases a contraction of the discharge section area appears. Then the effective volume of water flows through the opening at greater velocities than in a free channel.

It is necessary to determine the following elements by computation:

1. *Required discharge section area under the bridge—Contraction allowed.* The required smallest discharge section area under the bridge can be

determined by several methods. In the following, we will analyze the method based on average velocity in the main channel.

Using this method, the required smallest discharge section area under the bridge is computed from the formula

$$A = \frac{Q}{\mu V_0} ft^2 \tag{4.3}$$

where

A = the minimum section area required to allow the discharge to pass under the bridge considering contraction

Q = discharge with adopted probability of occurrence for the entire cross-section of channel and valley flats (ft^3/sec)

V_0 = average velocity in the main channel before the building of a bridge for a discharge with adoped probability (ft/sec)

μ = coefficient of contraction

If the bridge opening is located at an angle ϕ to the direction of the stream at a water stage adopted in computation, the discharge section area A under the bridge should be increased by multiplying it by sec ϕ.

2. *The coefficient of contraction.* Water does not flow under the bridge through the full cross-section, because a part of this cross-section is occupied by the bridge piers. Near the piers water flows less intensively; the eddies formed in these areas are caused by the insufficiently streamlined shapes of the piers and the possible roughness of the underwater surface of the piers. For this reason, the discharge area decreases under the bridge, a factor to be taken into account when computing by means of a coefficient of contraction. Formulas for computing the coefficient of contraction depending on the shape of piers and current velocity have been derived by Ozherelyev[11] using observations of discharges of high waters under bridges. The following two cases were considered:

a. For piers having the front part sharpened as an elongated triangle (Fig. 4.13):

$$\mu = \frac{L_c - 2(0.625V - 0.20)(n - 1)}{L_c} \tag{4.4}$$

b. For piers with a semicircular front part (Fig. 4.14):

$$\mu = \frac{L_c - 2(0.714V - 0.18)(n - 1)}{L_c} \tag{4.5}$$

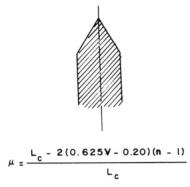

$$\mu = \frac{L_c - 2(0.625V - 0.20)(n - 1)}{L_c}$$

Figure 4.13 Pier with a sharpened front part.

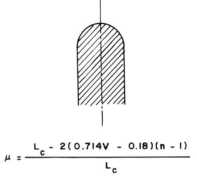

$$\mu = \frac{L_c - 2(0.714V - 0.18)(n - 1)}{L_c}$$

Figure 4.14 Pier with a semicircular front part.

where

L_c = the clear bridge span (ft)
n = number of bridge supports (piers and abutments)
V = water velocity adopted for computing the water discharge (ft/sec)

4.5 SCOUR

Scour is one of the basic characteristics of streams and is defined as the removal of stream bed material by stream or tidal currents. It may occur naturally as the result of channel construction or changes in the flow pattern. Great scour usually occurs during the larger floods. Scour may occur in the following cases:[12-15]

a. In the stream itself with or without a bridge.
b. At a bridge when stream flow is constricted by the bridge.
c. Local scour—due to alteration of the stream flow pattern by the bridge pier and abutment.

4.5.1 Local Scour

Local scour occurs around piers and abutments, is associated with vortex systems, and is due to the following:[16-18]

a. A bend in the channel (Fig. 4.15).
b. Piers shapes (Fig. 4.16).
c. Obstructions—the local scour hole around a bridge pier or abutment occurs because the pier or abutment is an obstruction to the flow, for example, at the skew bridge (Fig. 4.17).

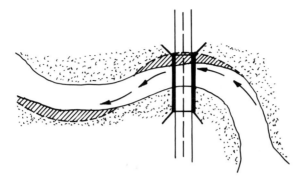

Figure 4.15a Bend in channel.

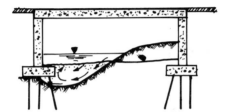

Figure 4.15b Scour action at bend.

4.5.2 Factors Affecting Scour

The depth and area of scour at a given bridge at a given time may be affected by any or all of the following factors:

- Slope, natural alignment, and shifting of the channel
- Type and amount of material in transport
- History of the former and recent floods
- Accumulation of ice, logs, or other debris
- Construction and/or realignment of flow due to the bridge and its approaches
- Geometry and alignment of piers
- Classification, stratification, and consolidation of bed and subbed material
- Placement or loss of rip-rap and other protective materials
- Natural or man-made changes in flow or sediment regimes
- Increase in stream discharge, increased velocity, and depth
- Temporary increase of discharge caused by flood, snow melting, or a gate opening
- Increased sediment load changes

78 CROSSING THE RIVER

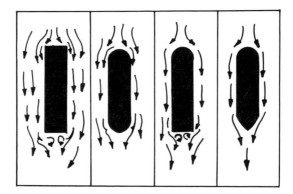

Figure 4.16a Shapes of the piers in plan.

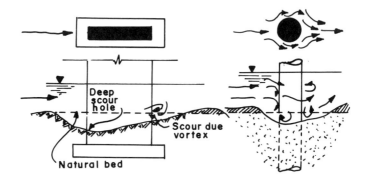

Figure 4.16b Local scour due to shape.

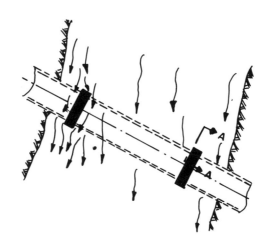

Figure 4.17 Obstruction of pier at skew bridge.

- Obstruction in the path of flow: pier, abutment, or debris
- A skewed pier causing a large obstruction

4.5.3 Protection of Foundation Against Scour

The need for scour protection can be minimized by locating bridges on stable tangential reaches of channels and by placing foundations on inerodible materials. However, such a solution is not always practicable, economical, or desirable from a road alignment standpoint. Several other alternatives include (Fig. 4.18):

1. Placing the bottom of the pier footing below the estimated lowest scour level, making allowance for local scour caused by the pier shaft and footing, and including an appropriate margin of safety.

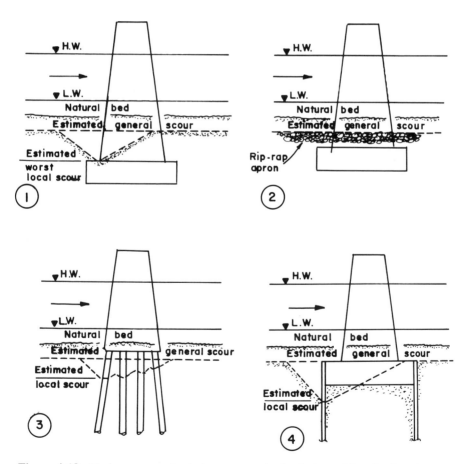

Figure 4.18 Various ways of catering to scour in the design of the pier foundation.

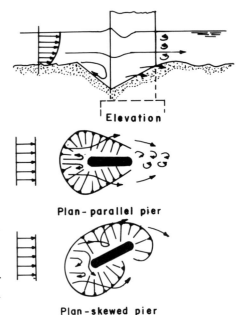

Figure 4.19 Usual form of local scour holes at piers, as demonstrated by model experiments.

2. Placing the bottom of the pier shaft below the estimated lowest general scour, and providing protection against local scour effects (Fig. 4.19).
3. Supporting the pier shaft or footing on piles or columns sunk well below the lowest scour level and designed to be secure when their upper parts are exposed by scour.
4. Constructing the pier in the form of a row of piles or columns without a footing or solid shaft, sinking these well below estimated scour levels and designing them to be secure when their upper parts are exposed by scour.
5. Supporting the pier as in (3), but protecting the upper parts of piles or columns against exposure by local scour.
6. Protecting the spread footing or piles against undermining by means of a sheet-pile skirting tied to the foundation. The skirting must be designed against scour and loss of support.

Selecting one alternative depends on a great many factors, including load-bearing requirements, subsoil conditions actually encountered, economics, feasible construction methods and schedules, and inspection procedures.

4.5.4 Suggestions to Minimize Scour Effects

The following are suggested to minimize scour effects in a new design:

1. Pier scour could be minimized by
   a. Alignment with stream flow

b. Placing the footing below the estimated scour depth, providing riprap at expected scour depth, or providing slope for the scour hole to allow the natural flow of water.
c. Using a pile foundation or caissons to provide stability when scour occurs
d. Providing sheet pile protection
2. Abutment scour could be reduced by
   a. Rip-rap protection
   b. A sheet pile
   c. A spur dike

4.5.5 Estimate of Possible Scour

In designing water-crossing bridges, an estimate of possible scour is required because scour may occur unless the stream bed is inerodible rock. The following three design factors should be considered:

1. The flood frequencies and magnitude
2. The flow pattern for each flood with geometry of the given design
3. The resulting scour: if a scour prediction cannot be made explicitly, the foundation should be designed more conservatively to accommodate unexpected scour depth

4.5.6 Computation of the Bridge Opening When Scour or Erosion Is Not Permitted

Erosion of the bottom of the river cannot be permitted when the bridge foundation is on piles. As (4.3) indicates,

$$A = \frac{Q}{\mu V_0}$$

The actual discharge section area under the bridge, after subtracting the area of the piers, is always smaller than the required discharge section area of the river before bridge construction. This may result in an excessively long bridge and it will be necessary to reduce this length. The reduction may be achieved either by deepening the river cross-section under the bridge or by permitting an increase in the velocity of the current.

1. Deepening the river cross-section may be achieved by cutting the banks or the valley flats to the average water level. This cutting is performed upstream and downstream of the bridge at a length of not less than 150 ft. The deepening should be such that the obtained cross-section is not less than $A = Q/\mu V_0$.

2. When we allow the current velocity under the bridge to be increased, by comparison with the velocity of the current before bridge construction, it is necessary to provide rivetment of the banks and bottom of the river. Bank rivetment is achieved by sloping the banks to make them stable, protecting them from scour by brush mattresses, rip-rap block paving, timber planning, piling, and sheet piling. The bottom of the river is protected against scour by rip-rap. The discharge section area under the bridge in the case of the artificial rivetment of the banks and river bottom may be determined as follows.[19,20] Let us designate:

V_1 = the velocity of the current at the riveted cross section
V_2 = the velocity of the current in the original cross-section
A_1 = the discharge section area after riveting
A_2 = the original cross section

In both cases the discharges should be equal, therefore,

$$A_1 V_1 = A_2 V_2 \quad \text{and} \quad A_1 = \frac{V_2}{V_1} A_2; \quad V_1 = V_2 \frac{A_2}{A_1} \qquad (4.6)$$

3. When scour is permitted, we adopt a greater velocity under the bridge than in a free channel.

Let us consider cross sections of a river before and after a bridge is built (Figs. 4.20 and 4.21).

We use the following designations:

A = The minimum section area required to allow the discharge quantity to pass under the bridge considering contraction, before building the bridge (ft^2)

A_a = The actual discharge section area existing under the bridge, after subtracting the area of the wet piers, considering the clear bridge span scour occurs (ft^2)

A_p = Wet area of the piers before scour

$$L_c = b_1 + b_2 + b_3 + b_4 = \text{the actual clear span} \qquad (4.7)$$

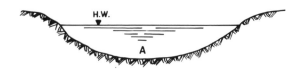

Figure 4.20 Original discharge cross section of the river.

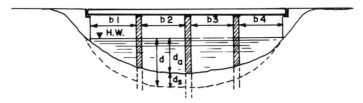

Figure 4.21 Cross section of the river after bridge construction.

The actual discharge section or active area is

$$A_a = A = A_p \quad \text{or} \quad A - A_a = A_p \tag{4.8}$$

Scour cross-sectional area should be equal to the wet area of the piers or

$$d_s(b_1 + b_2 + b_3 + b_4) = d_s L_c = A_p = A - A_a \tag{4.9}$$

and the average scour depth is in feet,

$$d_s = \frac{A - A_{a?}}{L_c} \tag{4.10}$$

Before scour the average depth of the river was

$$d_{\text{aver}} \, xL_c = A_a \tag{4.11}$$

or

$$d_{\text{aver}} = \frac{A_a}{L_c} \tag{4.12}$$

After scour this depth will increase; therefore, the total depth of the river bottom is

$$d = d_{\text{aver}} + d_s = \frac{A_a}{L_c} + \frac{A - A_a}{L_c} = \frac{A}{L_c} \tag{4.13}$$

The ratio between the total depth after scour and before scour we denote as

$$C_s = \frac{d}{d_{\text{aver}}} = \frac{A}{L_c} : \frac{A_a}{L_c} = \frac{A}{A_a} \tag{4.14}$$

and call this the coefficient of scour.

We consider that the bottom of the river at every section keeps the same ratio C_s between the depths of the water before and after the scour. Therefore,

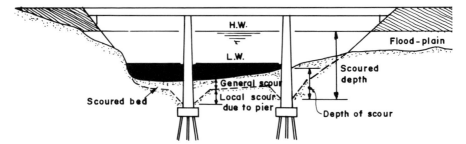

Figure 4.22 Cross section of a river under a bridge with scour depth indicated.

the depth after scour will be

$$d = C_s d_{\text{aver}} \tag{4.15}$$

Considering the maximum depth d after scour, we may determine the lowest elevation of the bottom of the river after scour and this elevation indicates the top of the pier foundation which should be provided in the design. It has been established by studying the discharge section area under existing bridges that in practice the coefficient of scour C_s fluctuates mostly within limits of 1.2 and 1.4, and very seldom reaches 1.5.

The bottom configuration line of the scour is obtained by plotting the depths computed by using Eq. (4.15),

$$d = C_s d_{\text{aver}}$$

at each depth of the cross section under the bridge from high-water level (Fig. 4.22). The area of a cross section under the bridge limited by the water surface at the water stage adopted in computations, and the line of scour, should be equal to the discharge section area under the bridge.

REFERENCES

1. Linsey, R. K. Jr., Kohler, M. A., and Paulhus, J. L. H. *Applied Hydrology*, McGraw-Hill, New York, 1949, pp. 243–249.
2. Neil, C. R. (Ed.), *Guide to Bridge Hydraulics*, University of Toronto Press, Toronto, 1973, pp. 13–46.
3. Sokolov, M. L., *Investigation of Bridge Crossing*, Avtotransizdat, Moscow, 1959, pp. 29–40 (in Russian).
4. Boldakov, E. V., *Crossings Over Small Watercourses*, Dorizdat Moscow, 1950, pp. 201–204 (in Russian).
5. Andreev, O. V., *Design of Bridge Crossing*, 2nd ed., Avtotransizdat, Moscow, 1960 (in Russian).

6. Boldakov, E. V. and Andreev, O. V., *Crossings Over Watercourses*, Avtotransizdat, Moscow, 1956, pp. 190-227 (in Russian).
7. Waddell, J. A. L., *Bridge Engineering*, vol. II, Wiley, New York, 1916, pp. 1109-1136.
8. Jarocki, W., *Computation of Bridge Spans and Culverts*, published by the Department of the Interior pursuant to an Agreement with the National Science Foundation, Washington, D.C., by the Scientific Publications Foreign Corporation Center of Technical and Economic Information, Warsaw, Poland, 1964, pp. 242-245.
9. Jarocki, W., op. cit, pp. 246-248.
10. Boldakov, E. V., *Crossings of the Large Rivers*, Dorizdat, Moscow, 1949, pp. 143-165, (in Russian).
11. Lebedev, V. V., *Hydrological Investigations and Calculation During Design of Bridges and Culverts*, Gidrometeorologicheskoe Izdateljstvo, Leningrad, 1949, pp. 201-203, (in Russian).
12. Kuhn, S. H. and Williams, A. A. B., Scour depth and soil profile determination in river beds, *Proceeding of the 5th International Conference on Soil Mechanics and Foundation Engineering*, 1961, pp. 487-490.
13. Laursen, E. M., Scour at bridge crossings, *Trans. ASCE*, **127**, part 1, 166-190 (1962).
14. Neill, C. R., op. cit, pp. 73-94.
15. Rotenburg, I. S., Voljnov, V. S. Polyakov, M. P., *Bridge Crossings*, published by High School, Moscow, 1977, pp. 182-212 (in Russian).
16. Neill, C. R., op. cit., pp. 94-109.
17. Rotenburg, I. S., Voljnov, V. S. Polyakov, M. P., op. cit., pp. 151-172, 212-228.
18. Neill, C. R., op. cit., pp. 110-141.
19. Jarocki, W., op. cit., pp. 283-289.
20. Lebeder, V. V., op. cit., pp. 204-206.

CHAPTER 5

STRUCTURAL BRIDGE SYSTEM

5.1 SUPERSTRUCTURE AND SUBSTRUCTURE

Every bridge structure consists of two basic parts:[1] superstructure (Fig. 5.1a) and substructure (Fig. 5.1b and c). The superstructure serves to take traffic loads and transfer them to the substructure, which generally consists of the piers (Fig. 5.1c) and abutments (Fig. 5.1b). The superstructure consists of the span between supports which carry the highway or railway and transfers this load to the substructure. Substructure takes the load and transfers it to the ground. Superstructure basically consists of the following parts.[2,3]

1. Elements that transfer the traffic load along the span onto the substructure, generally parallel to the longitudinal bridge axis. These elements are called the main carrying bridge members. Because the main bridge loading is vertical, the main carrying bridge members are vertical. These members, in the shape of plate girders, deflect under the loading and the resulting stresses are taken by the flanges. The webs of the plate girders and diagonals of trusses take shear forces.

2. Elements that transfer pressures from the vertical loads to the main carrying members in the transverse direction normal to the bridge axis, and connecting main carrying members in the transverse direction, are called the deck and transverse bracings or transverse construction (Fig. 5.2). This transverse construction is necessary because the main carrying members, installed as the plane walls, are placed at a certain distance from each other and, therefore, cannot take pressure from the loads that are placed between them. Apart from

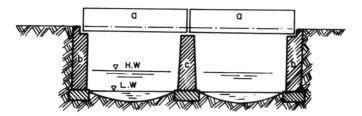

Figure 5.1 Superstructure and substructure of the bridge.

this, main carrying members as plane structures are unstable without a transverse connection.

3. Elements that transfer to the supports load resulting from the wind and centrifugal force. These loads are horizontal and the elements transferring these loads are located in horizontal planes, usually at the planes of flanges of the main carrying members. They are called the wind bracings because the main load acting on them is the wind. They are also called the transverse bracings because they are working in the transverse direction when they transfer wind loading to the supports. The main carrying members together with the deck and bracings constitute a superstructure unit. It is generally rectangular; its vertical sides are the main carrying members and wind bracing that corresponds to the distribution of the main and secondary loading—vertical and horizontal—is located along the horizontal side.

Traffic along the span may be arranged either above the top chord, between the main carrying members, or above the bottom chord, also between the main carrying members (Fig. 5.3). In this case the system consists of the vertical main carrying members with transverse beams and horizontal bracing arranged between the main carrying members in the plane normal to the bridge axis.

The position of transverse bracings at intermediate cross-sections of the bridge is very useful structurally because, by connecting the main carrying members, they reduce their relative overloading. When the deck position is at the bottom chord and main carrying members are not high, traffic prevents fully developed transverse bracing. In this case, the bridge in cross-section is open and is called a half-through span. Its free upper chords are stiffened in the transverse direction by special diaphragms (Fig. 5.4).

The open chord is obviously less stable than the chord that is stiffened by the wind bracing. For this reason, the clearance height of the bridge should not obstruct the position of the top bracing between the top chords, and the

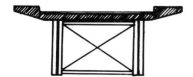

Figure 5.2 Cross-section of the bridge deck and transverse bracing.

88 STRUCTURAL BRIDGE SYSTEM

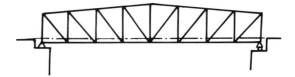

Figure 5.3 Deck at the bottom chord.

bridge is therefore closed (Fig. 5.5). With large spans the height of the bridge structure is usually greater than the clearance height. If sufficient clearance exists it is desirable to install transverse bracing (Fig. 5.6). End entrances to the bridge having a deck at the bottom chord cannot have bracings because of the required clearances. For this reason they are substituted by a system of rigid frames, called support frames or portal frames (Fig. 5.7).

The bridge superstructure is supported by the bearings. Bearings transfer the weight of the superstructure and traffic loading to the supports at definite locations. The intermediate supports are piers and the end supports are abutments. The intermediate supports have the shape of columns; in the cross-section they are configured such that the water will not produce whirlpool and scour. The abutments take the end reactions from the superstructure and also act as retaining walls. Both piers and abutments are treated in detail in Chapters 8 and 9, respectively.

5.2 BRIDGE GEOMETRY

The basic or general dimensions of the bridge are as follows:[4,5]

1. *The opening of a bridge.* L_0 is the distance between the front walls of

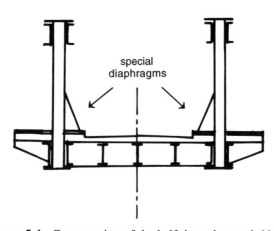

Figure 5.4 Cross-section of the half-through truss bridge.

5.2 BRIDGE GEOMETRY 89

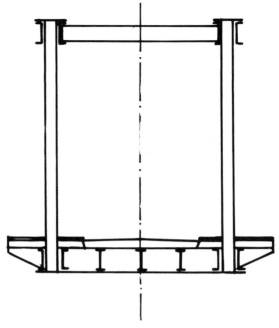

Figure 5.5 Cross-section of the through-truss bridge.

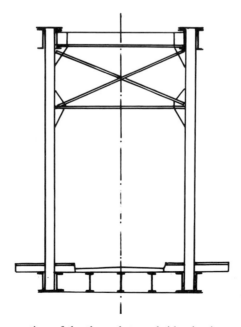

Figure 5.6 Cross-section of the through-truss bridge having transverse bracing.

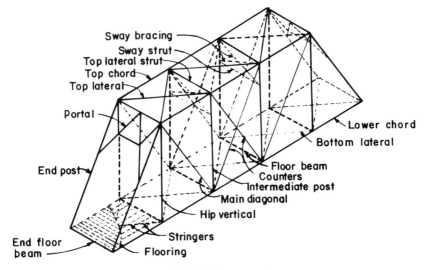

Figure 5.7 End support frames.

abutments opening (Fig. 5.8). Therefore the opening of the bridge is

$$L_0 = b_1 + b_2 + b_3 + \cdots + a_1 + a_2 \tag{5.1}$$

2. *The clear span.* The distance L_0 less the total width of all the bridge piers is called the clear span of the bridge (Fig. 5.8) or

$$L_c = L_0 - a_1 - a_2 \cdots = b_1 + b_2$$

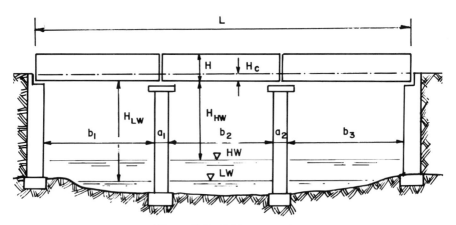

Figure 5.8 Bridge geometry.

where

b_1, b_2, b_3 = the individual clear spans
a_1, a_2, a_3 = the widths of the piers at the elevation of high water

3. *The length of the bridge.* The length of the bridge is the length of the superstructure of the bridge in elevation (Fig. 5.8).

4. *The span of the bridge.* The span of the bridge is the distance between center axis of the bearings in elevation (Fig. 5.8).

5. *The width of the bridge.* The width of the bridge is measured normal to the bridge as in Figure 5.5.

6. *Height of the main carrying members, namely, plate girder or truss.* The height of the plate girder or truss H is measured from the top flange of the top chord to the bottom flange of the bottom chord (Fig. 5.8).

7. *Construction height of the bridge.* The construction height of the bridge H_c is measured from the top of the bridge deck to the bottom flange of the bottom chord (Fig. 5.8).

8. *The distance from the top of the deck to the surface low water.* This distance is denoted by H_{LW}.

9. *Clearance under the bridge, or the distance from the bottom chord of the bridge to the level of the high water.* This distance is denoted by H_{HW}.

The values are basic because they characterize all particulars of the project and should be defined before starting the design. When the bridge is crossing a highway, the opening is determined by the volume needs of the traffic passing on the highway below. If the bridge is crossing a river, the opening is determined by the quantity of water passing under the bridge and by shipping conditions. At a gorge crossing the opening is determined on the basis of economical considerations and whether it would be better to substitute a high embankment in this situation.

In the case of a highway crossing, the opening is also determined by the width of the roadway on the bridge. In the case of a river crossing, the opening should be large enough that the maximum discharge of the river flow, which is determined by observation or by empirical formulas, may pass under the bridge at a velocity that does not cause the piers to become unstable, or cause scour of the river channel, or difficulties for shipping, and also at which the rising of the river level before the bridge is not too high.

Navigation channels are usually crossed by single-span bridges. When convenient conditions for crossing the river do not exist, it is necessary to design greater spans, as required by the shipping conditions. The spans of bridges in large cities may be established according to the local conditions and separately in each case. When the span is not determined by the shipping conditions, it is determined by the economic conditions, namely, by establishing the minimal cost of the bridge. Also, cost should be analyzed when a small span is recommended because of shipping conditions. It may turn out, particularly at small spans, that the most economical span is actually larger than that determined by shipping conditions.

5.3 DETERMINING THE MOST ECONOMICAL BRIDGE SPAN

The determination of the bridge span based on economic principles may be made as follows:[6]

Apart from shipping requirements, economic constraints are the most important factors to consider when crossing the bridge scheme and the sizes of separate spans, because span size substantially influences the cost of the bridge. We will consider bridge cost when different spans are used. We assume that geological conditions at the bridge location are similar. Then the intermediate piers may have equal volumes. The quantity of material for one pier may be expressed approximately by the formula

$$V_1 = V_0 + \beta l \qquad (5.2)$$

where

V_0 = that part of the pier volume that does not depend on the span of the bridge

βl = express dependence of the pier volume from the span l of the bridge.

Assuming that volume of piers V_2 does not depend from span values, we may express the cost W_1 of all bridge piers by the formula

$$W_1 = \left(\frac{L}{l} - 1\right)(V_0 + \beta l)k_2 + 2V_2 k_2 \qquad (5.3)$$

where

$\left(\dfrac{L}{l} - 1\right)$ = the number of the intermediate piers

k_2 = cost of the unit volume of the pier.

The cost of materials for carrying spans consists of the volume for the main constructions (beams, trusses, arches) and material for the deck. The cost of material per foot of deck does not depend on the span. The cost of the material per foot of the main carrying elements of the span is approximately proportional to the span. Therefore, the total amount of material per foot of the span may be expressed using the following variables:

$$g = g_{\text{deck}} + \alpha l$$

where

g = linear cost of the material for span l
g_{deck} = cost of material per foot of the deck
αl = cost of material per foot of the main carrying construction of the span.
α = coefficient depending on the span, but usually taken as a constant for small changes of the span. The value of this coefficient may be obtained from the data of earlier completed projects.

The cost W_2 of all spans of the bridge having total length L is:

$$W_2 = g_{\text{deck}} + \alpha l \, L k_1 \tag{5.4}$$

where k_1 is the unit volume (weight) cost of the span material.

The total cost of the bridge W will be

$$W = W_1 + W_2 = \left(\frac{L}{l} - 1\right)(V_0 + \beta l)k_2 + 2V_2 k_2 + (g_{\text{deck}} + \alpha l)L k_1 \tag{5.5}$$

We find the value of span l, when the cost of the bridge will be minimum after differentiating this expression by l and equalizing to zero, or

$$\frac{dW}{dl} = -\frac{LV_0 k_2}{l^2} - \beta k_2 + \alpha L k_1 = 0 \tag{5.6}$$

From this equation it is possible to determine the value of the optimum span, namely

$$l_{\text{opt}} = \sqrt{\frac{V_0 k_2}{\alpha k_1 - \dfrac{\beta k_2}{L}}} \tag{5.7}$$

In bridges having spans of beam systems the cost of material for intermediate piers basically depends on bridge width, height of the pier, geological conditions and, to a smaller degree, the value of the bridge span. If we use $\beta = 0$ and $V_0 = V_1$, then from equation (5.7) it is possible to determine the value of the optimum span as:

$$l_{\text{opt}} = \sqrt{\frac{V_1 k_2}{\alpha k_1}} \tag{5.8}$$

From this formula we obtain

$$\alpha l^2 k_1 = V_1 k_2 \tag{5.9}$$

and because αl expresses the expenditure of material per foot of the main carrying members of the structure, then $\alpha l^2 k_1$ represents the cost of a single span without a deck.

Therefore, (5.9) indicates that at the optimum layout the cost of one span without a deck should be equal to the cost of the pier. Comparing (5.9) and (5.8) indicates that the optimum span for bridges having thrust, defined by (5.7), is always greater than that for beam bridges.

When the configuration of the river channel or depth of pier foundation, which depends on geological conditions, abruptly changes along the length of the bridge, the method given above is not useful for determining the optimum span. If the river channel has a row of characteristic sections, and along each of them the conditions of building piers is approximately equal, the problem of laying out spans may be solved by determining the optimum span for each section separately. If the bridge crossing is impossible to divide into sections, and the dimensions of the piers are different at different locations, the optimum spans along the bridge length theoretically will also be different.

The theoretical data and formulas given above for determining optimum layout of spans are based on certain simplified assumptions. For this reason

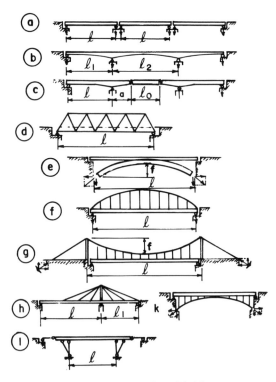

Figure 5.9 Types of steel bridges.

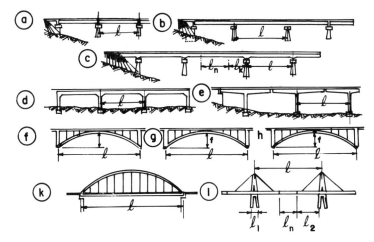

Figure 5.10 Types of concrete bridges.

they may be used only as a starting point. The final layout must be determined by comparing different alternatives and their technical-economical feasibility.

5.4 BRIDGE TYPES

In bridge engineering there are several different types of steel bridges[7] (Fig. 5.9) and concrete bridges[8] (Fig. 5.10). However, there are actually only three basic types of bridges, depending on the form of the load-bearing structure and its shape, which can be flat, convex, or concave. Designers think of them as beam, arch, and suspension or cable-stayed bridges (Fig. 5.11).

Figure 5.11 shows clearly the essential technical differences between these three types, that is, the direction of the forces they extend on their foundation. The beam is horizontally self-supporting. It exerts mainly vertical downward thrust on its piers. However, wind and braking forces may produce considerable horizontal forces. The support of an arch tends to open under the load, adding an outward horizontal component to the downward and inward thrust at its supports.

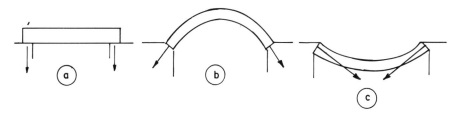

Figure 5.11 Concepts of the (*a*) beam, (*b*) arch, and (*c*) suspension bridges.

REFERENCES

1. Streletzkii, N. S., *Course of Bridges*, Moscow, 1925, p. 8 Published by High Technical College (in Russian).
2. Tall, L., (Ed.), *Structural Steel Design*, 2nd ed., Robert E. Krieger Publishing, Malabar, FL, 1983, pp. 94–143.
3. Kuzmanovic, B. and Willems, N., *Steel Design for Structural Engineers*, 2nd ed., Prentice-Hall, Englewood Cliffs, NJ, 1983, pp. 1–52.
4. Jarocki, W., *Computation of Bridge Spans and Culverts*, published for the Department of the Interior pursuant to an Agreement with the National Science Foundation, Washington, D.C., by the Scientific Publications Foreign Cooperation Center of the Central Institute for Scientific Technical and Economic Information, Warsaw, Poland, 1964, pp. 283–285.
5. Kolokolov, N. M., Kopac, L. N., and Fainstein, I. S., *Artifical Structuers*, 3rd ed., Transport, Moscow, 1988, pp. 18, 19 (in Russian).
6. Gibshman, E. E. (Ed.), *Bridges and Structures on the Highways*, Vol. 1, Published by Transport, Moscow, 1972, pp. 21–23 (in Russian).
7. Gibshman, E. E., *Bridges and Structures on the Highways*, Vol. 2, Published by Transport, Moscow, 1972, p. 22 (in Russian).
8. Gibshman, E. E., Vol. 1, op. cit., p. 203.

CHAPTER 6

SUPERSTRUCTURE—STEEL BRIDGES

6.1 ROLLED-BEAM BRIDGES

There are two main types of beam bridges—the simple span beam supported at its ends and the cantilever, or a beam which substantially overhangs its main supports. There are variations of both kinds of beam; the most common is the truss, generally a combination of linked triangles or other various configurations. While beam bridges of both types exert a vertical downward thrust on their supports, the cantilever, owing to its inherent tendency to pivot when the overhang is loaded, exerts an additional upward thrust at the other end.[1-5] The piers for both beam types normally have to support vertical load only, and are therefore comparatively simple in design. But the forces within a beam vary in its different parts, and include both thrust and tension. The material used in a beam bridge must be capable of withstanding both tension and compression.

The simple span beam is found, in practice, to provide the most economical form of bridge when the span is relatively small, namely up to about 150 ft (45 m). Where a large gap is to be bridged and piers will not be too costly, a multispan beam bridge is often the engineer's solution. The cantilever provides a means to make beam bridges of considerably greater spans, and there was a period in the history of bridge engineering when the steel cantilever only was in fashion and held the long-span record. Today the cantilever has many forms and uses, but long cantilever spans do not often appear to meet contemporary requirements. The arch avoids the complications of girder spans in tension and can bridge wider gaps. There are situations when a long continuous beam bridge is economical.

The beam bridge, in which a concrete roadway slab is supported by W section beams, is widely used because of its simple design and construction.

Rolled beams serve also as floor beams and stringers for the decks of plate girder and truss bridges. This type is economical for highway spans up to 80 ft and for railway spans up to 50 ft. Reduction in steel weight may be obtained by adding cover plates in the area of maximum moments, by providing continuity over several spans and designing the deck as a composite, or by a combination of these measures.

According to the AASHTO (American Association of State Highway and Transportation Officials) specifications minimum depth-span ratio is 1/25, and AREA (American Railway Engineering Association) recommends 1/15. These values can be extended somewhat by using cover plates or limiting the deflections for smaller depth-span ratios. Simple-span composite highway spans have been economically used for spans as large as 120 ft.

6.2 PLATE-GIRDER BRIDGES

The plate girder is a built-up beam consisting of two flange plates welded to a web plate to form an I (Fig. 6.1a) and box girders (Fig. 6.1b). Plate girders are used as primary supporting elements, as simple or continuous beams, and as stiffening girders of arch suspension and cable-stayed bridges. They also serve as floor beams and stringers on these and other bridge systems. Whereas rolled beams are available only in standard sizes, plate girders can be built to any desired size. In general, for smaller beams where the saving in material is minor compared with the increase in fabrication cost, rolled beams are more economical.

For heavier construction, where rolled beams are not sufficient to carry the load, plate girders are used. For modern highway bridges, where continuous and cantilever schemes are used to reduce the maximum moments, plate girders may not be required until the span exceeds about 80 ft because rolled

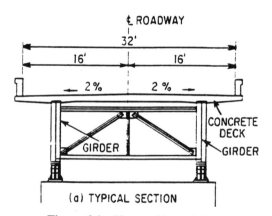

Figure 6.1 Plate and box girders.

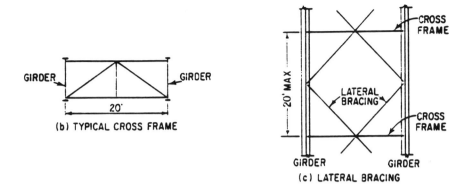

Figure 6.2 Lateral bracing and typical cross frame.

beams with reinforcing plates can often be used economically for such designs.[6-12]

Today plate-girder bridges are used for spans up to about 150 ft as well as for many larger spans. They are actually common for 200-ft spans and have been used for spans of well over 400 ft. The two or more girders supporting each span must be braced against each other to provide stability against overturning and flange buckling, to resist transverse forces, such as wind, earthquake, and centrifugal force, and to distribute concentrated vertical loads. On deck girders this is performed by a system of diagonals in the planes of the top and bottom flanges and by transverse bracing in vertical plane (Fig. 6.2).

The top lateral system can be omitted if the deck and its connections to the girders are designed to take its place. But a bottom lateral system is required for deck plate girders with spans greater than 125 ft. Transverse bracing should be installed over each bearing and at intermediate locations not over 25 ft apart. This bracing may consist either of full-depth cross frames (Fig. 6.2b) or of solid diaphragms at one-third or one-half the girder depth. Girder webs are protected against buckling by transverse and, in the case of deep webs, longitudinal stiffeners.

6.3 COMPOSITE BEAM AND GIRDER BRIDGES

The composite beam and girder construction for bridges defines a system in which the interaction of a concrete slab with a steel beam or girder is accomplished by means of a mechanical device called a shear connector.[13-15] The concrete slab becomes the compression flange of the composite beam or girder and the steel section resists the tension stresses. The tension portion of the beam or girder is usually not encased. The shear connector is usually in the form of studs, and may be as channels or spirals which transfer the longitudinal shear from the concrete to the steel and also hold the concrete from uplifting.

The use of composite beams or girders results in a cost savings because it makes a more efficient structure. In composite beam and girder bridges the resulting increase in effective depth of the total section and possible reduction of the top-flange steel usually allows some savings in steel compared with the noncomposite steel section. If the steel girder is not shored up while the deck concrete is placed, computation of dead-load stresses must be based on the steel section alone. Shear connectors should be capable of resisting all forces tending to separate concrete and steel surfaces, both horizontally and vertically.

6.4 CONTINUOUS COMPOSITE-PLATE GIRDER BRIDGES

6.4.1 Introduction

Bridges were designed for many years with each component having only one primary function. Gradually the structure has come to be viewed as a composite system. Considerable investigation went into the effort of evaluating the composite action of members. The use of composite construction lagged far behind experimental knowledge and was used only in a few isolated projects prior to 1940. The real impetus to composite construction in the United States came with the adoption of 1944 AASHTO specifications. Composite design rapidly became accepted for highway bridges.

Composite beams and girders are comprised of a steel beam and reinforced concrete slab interconnected so that the component elements act together as a unit.[16] The steel beam may be fully encased in concrete. The natural bond between the steel and the concrete can be relied on to provide composite action. Otherwise the composite action is assured by small pieces of steel bars or shapes welded to the top flange of the steel beam and embedded in the concrete of the slab (Fig. 6.3). The steel beam may be a rolled beam, a rolled beam with cover plates, or a built-up section. An asymmetrical section, such as a rolled beam with a cover plate on the bottom flange, is often economical. The reinforced-concrete slab acts as a very effective cover plate if it is on the compression side of the steel beam. Its dimensions are usually dictated by the

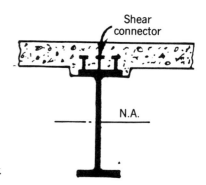

Figure 6.3 Composite beam.

beam spacing and the required capacity for load transfer to the beams. The design of the slab is independent of the composite action and is carried out in the same manner as for noncomposite floors.

The shear connectors or the concrete encasement provides the necessary connection between the slab and the beam. Their function is to transfer horizontal shear from the slab to the beams and to force the concrete and steel parts to act as a unit. In general the advantages of composite construction are:

a. A saving in the weight of steel, ranging from 20 to 30%
b. Reduction in the depth of flexural members
c. Increased stiffness of a floor system
d. Increased overload capacity of a floor

The relative economy of composite rolled beams versus noncomposite ones for use in highway bridges consisting of a simple-span structural system is shown in Figure 6.4.

6.5 OPTIMUM HEIGHT OF THE PLATE GIRDER

The optimum height of the plate girder may be obtained using a minimum of steel. The quantity of steel per foot of the girder may be expressed as the summary of two volumes:[17,18]

Expenditure of steel for chords of the girder

$$P_c = \frac{2S}{h} \psi_c \qquad (6.1)$$

Expenditure of steel for web

$$P_w = \delta h \psi_w = n h^2 \psi_w \qquad (6.2)$$

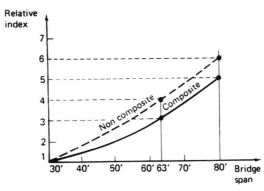

Figure 6.4 Relative economics of composite and noncomposite beams.

102 SUPERSTRUCTURE—STEEL BRIDGES

where

S = required section modules of the girder
h = height of the girder
δ = thickness of the web
$n = \delta/h = \frac{1}{100}$ to $\frac{1}{200}$ = ratio of the thickness of the web to its height
Ψ_c, ψ_w = structural coefficients of the chords and the web

The expenditure of steel for the girder will be the minimum, considering that the first derivative by h of the sum $(P_c + P_w)$ is equal to zero, or

$$\frac{d}{dh}\left(\frac{2S}{h}\psi_c + \delta h \psi_w\right) = -\frac{2S}{h^2}\psi_c + 2nh\psi_w = 0 \tag{6.3}$$

Then the optimum height of the girder is

$$h = \sqrt[3]{\frac{S\psi_c}{n\psi_w}} = \sqrt[3]{\frac{\psi_c}{n\psi_w}} \cdot \sqrt[3]{S} = m\sqrt[3]{S} \tag{6.4}$$

The value of the coefficient m is approximately equal to 6 or 7.
In modern simple-span plate girder bridges the height of the plate girder used is

$$\left(\frac{1}{10} \text{ to } \frac{1}{16}\right) 1$$

With increase of span the relative height of the bridge is made smaller and for continuous and cantilever girders sometimes is

$$\left(\frac{1}{20} \text{ to } \frac{1}{25}\right) 1 \text{ and smaller}$$

The quantity of the main girders in the cross-section of the bridge depends on the width of the roadway and span. Small-span spacing of main girders is small, from 6.5 to 9.8 ft (2 to 3 m). In this case the roadway may be directly supported by the main girders (Fig. 6.5a). With the increase in span it is convenient to use stronger, but fewer, main girders. For this reason, in bridges with large spans (Fig. 6.5b), when construction depth is not limited only two main girders are used in the cross-section of the bridge.
In the wide bridges (Fig. 6.5c) a greater number of main girders is used, installed at spacings of 20–26 ft (6–8 m). At greater spacings among the main girders the roadway is supported by the stringers: longitudinal (Fig. 6.5c), transverse (Fig. 6.5b), or transverse and longitudinal (Fig. 6.5d).

6.5 OPTIMUM HEIGHT OF THE PLATE GIRDER

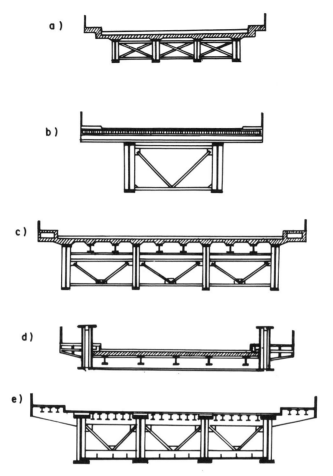

Figure 6.5 Placement of the main girders in the cross-section of steel span: (*a*) deck supported by main girders; (*b*) deck supported by two girders; (*c*) bridge having wide deck, greater number of main girders; (*d*) deck supported by the stringers and transverse beams; (*e*) orthotropic bridge.

In the case of limited construction depth, main girders are closely spaced, which permits a reduction in height.

Bridges having spans of 500–820 ft (150–250 m) require special construction shapes, which provide space stiffness of the span structure at minimum height. In this case, an orthotropic type bridge is used (Fig. 6.5*e*). In addition, composite plate girders are widely applied where a reinforced concrete beam takes compressive stresses and steel girders generally work with tension. Composite girders may provide 15–20% economy in steel and are most effective in simple spans. In composite girders reinforced concrete slab permits a great reduction in the cross-section of the upper flange of the plate girder.

6.6 HELPFUL HINTS FOR GIRDER BRIDGE DESIGN

The principal advantage of the rolled-beam bridge is its simplicity.[19,20] Fabrication is usually simple, no falsework is required, and erection is simple. Rolled beams are economical for spans up to about 60 ft. Plate girders have the advantage that they can be tailored to fit shear and moment requirements more closely than the rolled beam. They become advantageous at about 60 ft and are commonly used for spans to 300 ft or more. The deck bridge is used more often. Through-girder bridges are used where clearance is limited. The box girder is an efficient shape with superior torsional stiffness. It is suitable for long-span bridges, particularly when it is used in conjunction with an orthotropic steel plate deck. Composite construction is most economical for spans 60 ft and longer. Although continuous spans will often show economy in weight, requirements of field splicing, erection falsework, and equipment often outweigh savings in cost. A series of simple spans is usually better for bridges whose overall length is less than about 100 ft. However, many variables are involved in the decision, so that each location should be studied in light of conditions at the site.

Ordinarily, the number of beams or girders in a deck highway bridge should be such as to optimize the cost of the bridge. This involves a comparison of the cost of the beams or girders with that of the deck. For spans up to about 100 ft, either three or four girders will usually be economical for a two-lane highway bridge. As the span increases, the depth of the girder and its weight rise rapidly, so that for longer spans, two girders with a floor system will be economical.

For simple or continuous spans, deflection due to liveload plus impact should not exceed $1/800$ of the span. Deflection limitations often defeat the economy of higher-strength steels for short spans. The depth-to-span ratio of plate girders and rolled beams used as girders should be not less than $1:25$. For continuous spans, the span is taken as the distance between dead-load points of contraflexure.

A designer can obtain a fine design from a theoretical point of view by following specifications closely and using every advantage to save material. But this practice is not always the best design technique and may not yield the lowest-cost structure. The relationship between weight of materials and cost of structure depends on the answer to the question: Is it more economical to use more material and less labor, or more labor and less material? To produce lowest-cost structures that meet owners' requirements, designers need extensive and appropriate design experience, supplemented by knowledge of what is current good practice in fabrication. In addition, they should be familiar with good, economical erection procedures. Designers must keep abreast of new developments and add to their design capabilities accordingly.

In summary, a number of suggestions are offered to aid novice designers of beam and girder bridges:

- Take into account all three major elements—material, fabrication, and erection.

- Use the least number of stringers in a multibeam bridge that is compatible with a good deck design. A stringer omitted eliminates fabrication of such details as bearings, diaphragms, and connections and cuts erections costs.
- Save material by simplifying details. Use simple diaphragms, when they are adequate, instead of cross bracing. Use simple pins or rockers for bearing devices where they are applicable.
- Try to eliminate expansion joints wherever possible. They are expensive and, in most cases, hard to maintain. Lack of maintenance makes joints inoperative and could prevent proper functioning of the structure.

Under normal conditions, for shallow girders, it may be more economical to eliminate intermediate transverse stiffeners required for shear by utilizing a heavier web. A design with a heavier web may reduce fabrication costs significantly. For deep girders, web thickness should be selected to minimize the number of intermediate transverse stiffeners required. Also, consideration should be given to stopping intermediate stiffeners short of the tension flange, to reduce shop time in fitting up and, hence, costs.

In general, do not specify both longitudinally and transversely stiffened webs for girders unless large depth is required. When both types of stiffeners are used on the same web, they should preferably be placed on opposite sides of the web. This avoids interference problems and minimizes fabrication costs.

Continuous and suspended-span rolled beams and shallow girders normally do not require haunching. Fabrication usually is less costly for girders with parallel flanges. Whether girders should be haunched for appearance's sake is debatable. For girders of intermediate or large depth, however, there may be a real need to haunch because of clearance requirements. Otherwise, use of parallel flanges is advisable because of the cost savings that result from greater ease of fabrication.

During the design phase of deep girders, attention should be given to the problems of erection. Close collaboration with prospective erectors will often reveal items that, when incorporated in the design, will minimize difficulties in the field.

6.7 COMPOSITE BOX-GIRDER BRIDGES

Box girders are desirable for spans of about 120 ft and up.[21-23] Structural steel is employed at high efficiency in box girder bridges. Corrosion resistance is higher than in plate-girder and rolled-beam bridges. Also, the box shape is more effective in resisting torsion. In addition, box girders offer an attractive appearance (Fig. 6.6). A bridge may be supported on one or more box girders. Each girder may comprise one or more cells. For economy in long-span construction, the cells may be made wide (12 ft or more) and deep.

Box girders may be simply supported or continuous. They are adaptable to composite and orthotropic-plate construction. With composite construction, only a narrow top flange is needed with each web. Box girders may be simply

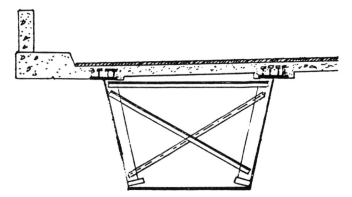

Figure 6.6 Box-girder bridge.

supported or continuous. They are adaptable to composite and orthotropic-plate construction. With composite construction, only a narrow top flange is needed with each web. Boxes may be rectangular or trapezoidal. Construction costs often are kept down by shop fabrication. Thus, designers should bear in mind the limitations placed by shipping clearances on the width of box girders as well as on length and depths. In orthotropic-plate girder construction, a steel-plate deck is used instead of concrete. The plate is topped with a wearing surface. The steel-plate deck serves the usual function of distributing loads to main carrying members, but it also acts as the top flange of these members. Orthotropic plates often are used with box girders.

6.8 ORTHOTROPIC DECK BRIDGES

The orthotropic bridge deck consists of a flat, thick steel plate, stiffened by a series of closely spaced longitudinal ribs of right angles, or othogonal, to the floor beams.[24,25] As the rigidities of the ribs and floor beams are generally of unequal magnitude, elastic behavior is different in each of these two main directions. Because ribs and floor beams are orthogonal and because in both directions their elastic properties are different or anisotropic, the whole system is known as orthogonal-anisotropic, or, briefly, orthotropic. There are two types of modern orthotropic deck—open-rib and closed-rib (Fig. 6.7).

Orthotropic steel plate bridges consist of orthotropic panels acting monolithically with a supporting grillage of beams and girders. When used on steel bridges, orthotropic decks are usually joined by welding or high-strength bolting to the main girders and floor beams. An orthotropic deck serves as a roadway and structurally as the top flange of the plate or box girders. The combination of plate or box girders with orthotropic decks permits the design of bridges of considerable slenderness and of nearly twice the span reached by girders with concrete decks. The orthotropic decks are widely used on contin-

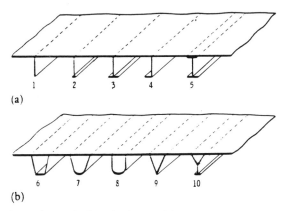

Figure 6.7 Orthotropic steel plate deck (*a*) with open ribs and (*b*) with closed ribs.

uous girders on low-level river crossings in metropolitan areas and for main spans of more than 1000 ft for suspension and cable-stayed bridges and up to 856 ft for plate-girder bridges.

It should be pointed out that orthotropic decks are generally not competitive, at least in the United States, owing to the large amount of fabrication and welding. It is generally used under special circumstances, such as, redecking an existing bridge "under traffic" with modular deck units (Golden Gate Bridge).

6.8.1 Box Girders

Box girders are often used to support orthotropic decks, especially when structural depth is restricted. In cross-section box girders are usually rectangular or trapezoidal. Wide decks are supported by either single box girder or twin boxes. Wide single boxes have been built with multiple webs or secondary interior trusses.

6.8.2 Cable-Supported Box Girders

Cable-supported box girder systems have either one or two box girders, with one or two towers for each girder. The cables are curved if the girders are suspended at each floor beam, otherwise they are straight. In the latter case, the cables either are inclined and parallel or converge from the towers.

6.8.3 Ribs

There are two basic systems of orthotropic steel decks used at present, characterized by torsionally soft or open ribs and torsionally stiff or box-shaped ribs (Fig. 6.8). As a rule, the ribs are arranged in the longitudinal direction of the bridge to provide maximum flexural rigidity against maximum moments acting in the longitudinal direction.

108 SUPERSTRUCTURE—STEEL BRIDGES

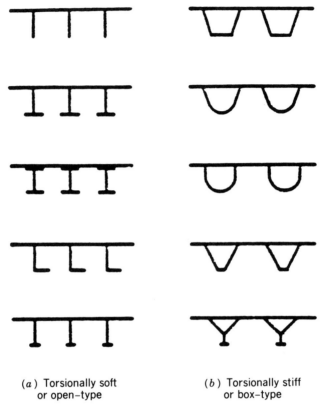

(a) Torsionally soft or open-type

(b) Torsionally stiff or box-type

Figure 6.8 Rib types.

The open ribs are usually made from flat bars, bulb shapes, inverted T-sections, angles, and channels. Because of the high stresses to be carried by these ribs, due to their participation in the girder action, they are made continuous through the slotted web plates of the floor beams and constructed of low-alloy structural steel. Open ribs usually vary in size as required from 3/8 in. × 8 in. to 1 in. × 12 in. along the cross-section of the bridge. The spans of the open ribs are in the range of 4 to 7 ft.

Among the many types of box-shaped ribs are trapezoidal, semicircular, triangular, and combined. The most often used is the trapezoidal rib section. This particular type possesses considerable torsional rigidity, resulting in a good transverse load distribution by the deck. Usually the box ribs continue through slots in the webs of the floor beams. The thickness of the ribs varies from 3/16 in. to 5/15 in.

6.8.4 Floor Beams

In the orthotropic deck, floor beams are usually made from steel plates welded together in the shape of an inverted T-section, and the top flange of the full

I-section is formed by the deck plate. Floor beams are usually spaced from 4 to 15 ft apart, but in exceptional cases this spacing may be increased. Special cutouts are usually provided in the floor beam webs at the intersections with the continuous longitudinal ribs.

6.8.5 Orthotropic Deck Surfacing

The wearing surface on a steel plate deck should satisfy the following requirements:

1. To reduce dead weight on the deck, the covering should be lightweight, but of sufficient thickness to cover all irregularities of the steel surface and thus provide an even, plane traveling surface.
2. To increase friction and minimize ice hazards, the surfacing should possess skid resistance.
3. Stability and durability should be maintained over the entire expected temperature range under the action of traction and braking forces of vehicles.
4. A good wearing surface should provide permanent corrosion protection, be impervious to water and chemical agents, develop no cracks through which moisture can penetrate onto the steel deck, and have good bondage at the surface of the steel under all conditions.

For various reasons, preference has been given to such bituminous-mix wearing surfaces as asphalt concrete and asphalt mastic. Usually, an appropriate insulating material is applied to the steel deck before the wearing surface. This protective layer provides positive corrosion protection for the steel deck and improves the bonding between the covering and steel plating.

6.9 TRUSS BRIDGES

Trusses are lattices formed of straight members in a triangular pattern.[26-29] Trusses are classified by the manner in which various systems of triangulation are combined. The top and bottom chords may be parallel or inclined, the truss may be simple span or continuous, and the end members may be vertical or inclined. A highway or railway truss may be defined by the manner in which it supports the load it carries, such as a deck truss, through truss, or half-through truss. The deck truss is built entirely below its load. The load passes between the trusses of a through bridge and below an overhead bracing system. A pony truss is a type of a half-through truss bridge which, because of its shallow depth, does not have an overhead bracing system.

The most common form of bridge truss is the Warren truss (Fig. 6.9a). As with most other trusses, the chords carry the bending moment and the diagonals carry the shear. The vertical members carry only the panel loads and can

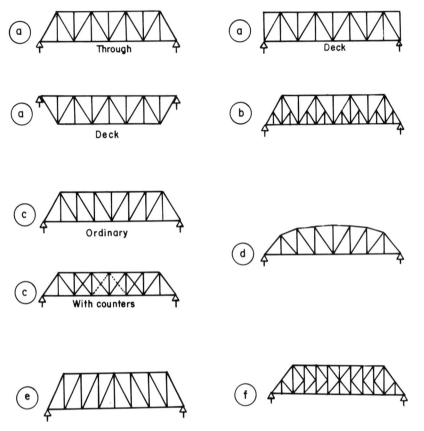

Figure 6.9 Types of bridge trusses: (*a*) Warren; (*b*) subdivided Warren; (*c*) Pratt; (*d*) curved-chord Pratt; (*e*) Howe; (*f*) K-type.

be economically designed. However, the Warren truss has relatively high secondary stresses, and a Pratt truss (Fig. 6.9*b*) is considered more desirable from that point of view.

Truss bridges require more field labor than comparable plate girders. Also, trusses are more costly to maintain because of the more complicated makeup of members and poor accessibility of the exposed steel surfaces. For these reasons, and as a result of changing aesthetic preferences, use of trusses is increasingly restricted to long-span bridges on which the relatively low weight is a decisive advantage.

The superstructure of a typical truss bridge is composed of two main trusses, the floor system, a top lateral system, a bottom lateral system, cross frames, and bearings. Decks for highway truss bridges are usually concrete slabs on steel framing. The economic height-to-span ratio for bridge trusses is about one-sixth to one-eighth, varying with the type of truss, loading, span length, and so on. It can be shown that the optimum inclination of the diagonals is about 45°. When truss spans are increased in length, their economical height will also increase. Thus, both the Warren and the Pratt trusses will result in long panel lengths if the diagonal inclination remains about 45°. One way to shorten the panel length—in order to shorten the span of stringers—is to subdivide these trusses. A subdivided Warren truss is shown in Figure 6.9c.

These subdivided trusses have the disadvantage of developing high secondary stresses, and K trusses may be preferred (Fig. 6.9d). The K trusses will keep the desirable diagonal inclination, supply the required depth of truss, and at the same time limit the span of the stringers.

Truss chords may be curved to carry part of the shear and to reduce stresses in the diagonals (Fig. 6.9e). There is a slight increase in the cost of fabrication compared to a parallel chord truss, but for medium and long spans the additional cost may be more than balanced by the saving in material. Sections of truss members are selected to insure effective use of material, simple details for connections, and accessibility in fabrication, erection, and maintenance. Preferably, they should be symmetrical. Welded truss members are formed of plates.

6.10 OPTIMUM HEIGHT OF THE TRUSS

The height of the truss in highway bridges may be determined considering minimum expenditure of the metal and requirements of the vertical stiffness of the bridge.[30] The optimum height of the truss when the expenditure of the metal is minimum may be determined as follows: The weight of metal of the truss is a combination of the weight of the chords and the web. The weight of the chords for a truss having constant depth may be expressed by the formula

$$W_c = \sum \frac{Md}{h\sigma} \gamma \psi_c \tag{6.5}$$

where

M = bending moment in the successive chord panels
h = height of the truss
d = length of the panel
σ = allowable stress for truss metal
γ = specific weight of the steel
ψ = construction coefficient of the chord indicating how much actual weight of structure elements is greater than its theoretical value

Similarly, it is possible to express the weight of the web. For trusses with diagonals their weight is

$$W_d = \sum \frac{Q}{\sigma \sin \alpha} s\gamma\psi_d \sum \frac{Q}{\sigma} h\gamma\psi_d \left[1 + \left(\frac{d}{h}\right)^2\right] \tag{6.6}$$

where

$$\frac{Q}{\sin \alpha} = \text{forces in diagonals}$$

$$S = \frac{h}{\sin \alpha} = \text{length of diagonal}$$

$$\frac{1}{\sin^2 \alpha} = 1 + ctg^2\alpha = 1 + \left(\frac{d}{h}\right)^2$$

The optimum height of the truss is obtained by taking the derivative from the total weight of the truss $W_c + W_d$ by h and equalizing to zero.

$$\frac{d(W_c + W_d)}{dh} = -\left[\sum \frac{Md}{\sigma}\gamma\psi_c + \sum \frac{Qd^2}{\sigma}\gamma\psi_d\right]\frac{1}{h^2} + \sum \frac{Q}{\sigma}\gamma\psi_d \quad (6.7)$$

From this expression we obtain the optimum height of the truss:

$$h_{opt} = d\sqrt{\frac{\sum \frac{M\phi_c}{d}}{\sum Q\phi_v} + 1} \quad (6.8)$$

Therefore, the optimum height of the truss depends on the panel and is increased with an increase in the panel length.

On the basis of experience in bridge design and considering requirements regarding stiffness, the heights of the trusses at highway bridges are used as follows:

For parallel chords

$$h = \left(\frac{1}{7} - \frac{1}{10}\right)l \quad (6.9)$$

For polygonal configuration

$$h = \left(\frac{1}{5.5} - \frac{1}{8}\right)l \quad (6.10)$$

In trusses having the web as triangle, the optimum length of the panel is $(0.6 - 0.8)h$ with verticals and $(1.0 - 1.2)h$ at web as triangle, where h is the height of the truss.

6.11 ARCH BRIDGES

6.11.1 General Characteristics of the Arches

Unlike the beam, which must withstand tensile as well as compressive forces, a true arch is in a state of compression throughout.[31-34] It can, therefore, be built of materials that, although strong in compression, are weak in tension. Arches are handsome in appearance and sometimes are used largely for that reason.

The arch is likely to continue to play a leading part in bridge construction but it has clear limitations because of the great headroom required as well as the high cost of the falsework and also the very exacting foundation requirements. A true arch becomes increasingly difficult to construct as the $h/1 =$ rise/span ratio decreases. The effects of the temperature fall and arch shortening due to the dead-load thrust induce large bending moments, particularly at the crown and springing, which may be greater than those due to live loads.

The flat arch is a very interesting structure and many attempts have been made to increase its economy by deriving a form of curve which would eliminate or reduce the effects of arch shortening. This is by no means easy. Improved methods of testing by means of large-scale models will undoubtedly lead to more accurate knowledge of the behavior and stability of flat arches and facilitate their use by engineers in special cases.

6.11.2 Basic Types of Arches

The three basic types of arches are shown in Figure 6.10. Rise ratio $= h/1$ is fixed largely to the topography, clearance, grade line, and appearance and varies from $\frac{1}{10}$ to $\frac{1}{2}$. Arches are classified as to the number of hinges used in one rib as:

1. Three-hinged, at crown and supports. Three-hinged arches possess the negligible influence of temperature on stresses.
2. Two-hinged, at supports. These arches have less deflection at the crown due to changes in temperature, and are more rigid than those having three hinges, but the temperature stresses are larger and any horizontal movement of one support produces stress throughout the arch.
3. A fixed arch has greater temperature stresses and is more affected by settlement of the supports.

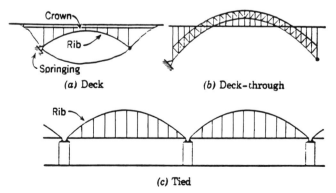

Figure 6.10 Basic types of arches.

6.11.3 Abutments

The abutments of arch bridges, except the tied arch, have to withstand the horizontal component of the arch's angular thrust as well as the vertical reactions. They are, therefore, of more complex design and consequently more costly. The low angle of angular thrust necessitates very large abutments unless there is suitable rock in the ground. Unlike the beam, however, which in itself must withstand tensile as well as compressive forces, a true arch is in a state of compression throughout. It can, therefore, be built of materials that, though strong in compression, are weak in tension.

The roadway of an arch bridge may be supported above the arch or suspended below it. The arch itself may be hinged which means it is pin mounted at its ends (the two-hinged arch) and possibly at the crown too (making a three-hinged arch); or it can be monolithic in its abutments, and, therefore, hingeless.

A true arch configuration may be circular or parabolic, or it can also be polygonal, built up of rigidly joined straight or nearly straight sections. The inherent features of the arch are its rigidity and strength. It is relatively easy to erect the arch where temporary falsework can be built. However, the long arch often confronts the engineer with unusual erection problems and must frequently be designed to withstand stresses during erection that it will never have to bear when complete.

A tied arch has its two ends connected by a horizontal member that, acting in tension, takes the full horizontal component of the arch's angular thrust, so that the force acting on the foundations is vertical only as in the case of simple beams.

6.11.4 Structural Characteristics

The inherent features of an arch are its rigidity and strength. It is easy to erect where temporary falsework can be built. But the situation that calls for an arch bridge means that intermediate piers are for some reason impractical—and often means that falsework is impractical too. In fact such situations often confront the engineer with unusual erection problems and the arch must be designed to withstand stresses during erection that it will never have to bear when completed.

6.11.5 Deck Construction

The roadway deck of steel arch bridges is usually of reinforced concrete, often of lightweight concrete, or a framing of steel floor beams and stringers. To avoid undesirable cooperation with the primary steel structure, concrete decks either are provided with appropriately spaced expansion joints or are prestressed. Orthotropic decks that combine the traffic deck, tie bar, stiffening girder, and lateral diaphragm have been used on some major arch bridges.

6.12 SUSPENSION BRIDGES

The three main supporting parts of a large modern suspension bridge are the cables, the towers, and the anchors (Figure 6.11). The cables pass over the saddle on top of the towers. Then the cables pass down to the anchors, which are usually a heavy mass of concrete. The cables pull upward and inward on these anchors. When the cables have been finished, the main supporting parts of the bridge are finished. The roadway is hung from the main cables by the vertical cable suspenders which are fastened at even spacings to the main cables, extending down to the road level. Then a truss or grider side may be built out from hanger to hanger.

6.12.1 Stiffening Girders or Trusses

The purpose of stiffening girders or trusses is to distribute concentrated loads, reduce load deflections, act as chords to the lateral systems, and secure the aerodynamic stability of the structure. Spacing is controlled by the roadway width but is usually not less than $\frac{1}{50}$ the span. The depth of stiffening trusses is at least $\frac{1}{180}$ of the main span. Panel length may be equal to, twice, or one-half the floor beam spacing, so the truss diagonals will be close to 45°.

6.12.2 Cable Systems

Sometimes, for greater areodynamic stability, the suspenders are interwoven with diagonals that originate at the towers. Main cables, suspenders, and stiffening girders or trusses are usually arranged in vertical planes, symmetrical

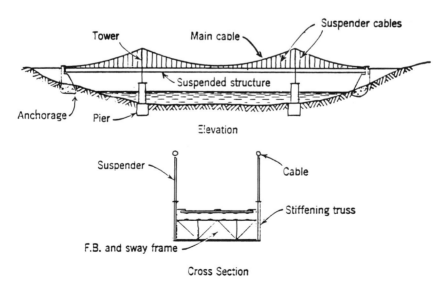

Figure 6.11 Typical suspension bridge.

116 SUPERSTRUCTURE—STEEL BRIDGES

with the longitudinal bridge axes. Three-dimensional stability is provided by top and bottom lateral systems and transverse frames, similar to those in conventional girder and truss bridges. Bridge roadway decks may take the place of either or both lateral systems.

The main cables are usually made up of galvanized bridge wires which are placed either parallel or in strands, and compacted and wrapped with small-gauge wire. Twin cables are used if larger sections are required. Suspenders may be rods or single steel ropes. Connections to the main cable are made with cable bands, which are clamped together with high-strength bolts.

6.12.3 Towers

The towers may be portal, multistory, or diagonally braced frame (Figure 6.12). They may be of cellular construction, made of steel plates and shapes, steel lattices, or reinforced concrete. The base of the steel tower may be fixed or hinged. The cable saddles at the top of fixed towers are sometimes placed on rollers to reduce the effect on the towers of unbalanced cable deflections.

6.12.4 Floor System

In the design of the floor system the main considerations should be the reduction of dead load and of resistance to vertical air currents. The deck is usually lightweight concrete, orthotropic deck, or steel grating partly filled with con-

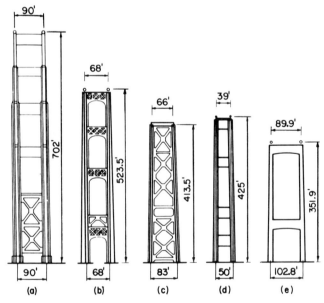

Figure 6.12 Types of towers. Reprinted with permission from Merritt, F. S., *Structural Steel Designers' Handbook*, McGraw Hill, New York, 1972.

crete. Expansion joints should be provided every 100–120 ft to prevent mutual interference of deck and main structure. Stringers should be made composite with the deck for greater strength and stiffness. Floor beams may be plate girders or trusses.

6.12.5 Continuity

The majority of suspension bridges have main cables draped over two towers. Therefore, such bridges consist of a main span and two side spans. The ratios of side span to main span are 1:4 to 1:2. The ratios of cable sag to main span are preferably in the range of 1:9 to 1:11. If the side spans are short enough, the main cables may drop directly from the tower tops to the anchorages. In that case, the deck is carried to the abutments on independent, single-span plate girders or trusses. When side spans are not suspended, the stiffening girder or truss is restricted to the main span.

6.12.6 Suspension Systems

Long-span suspension bridges are of two basic types—stiffened and unstiffened structures.[35-43] All important suspension bridges in clear spans up to 1800 ft were built with heavy stiffening trusses to prevent appreciable changes in shape of the parabolic cable curve under the effect of an unsymmetrically placed live load. It has been estimated that the unstiffened type could be used where the cable change under live load would not permit a slope or grade in excess of 5% at any point in the floor. However, in addition to the slope of the roadway, consideration must be given to the prossibility of wind-induced oscillation or galloping.

Several types of stiffened suspension bridges are illustrated in Figure 6.13. An important distinction rests upon the number of hinges in the stiffening truss, the three-hinged type being statically determinate, the two-hinged type having one redundant part, and the continuous type being three times statically indeterminate.

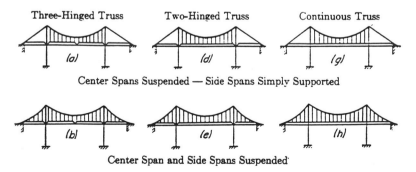

Figure 6.13 Types of suspension bridges.

6.13 CABLE-STAYED BRIDGES

During the past decade cable-stayed bridges have found wide application in many countries.[44-52] The renewal of the cable-stayed system in modern bridge engineering was due to the tendency of bridge engineers in Europe, primarily Germany, to obtain optimum structural performance from material that was in short supply during the postwar years. Cable-stayed bridges are constructed along a structural system that comprises an orthotropic deck and continuous girders that are supported by stays, that is, inclined cables passing over or attached to towers located as the main piers.

The application of inclined cables gave a new stimulus to the construction of large bridges. The renewed application of the cable-stayed system was possible only under the following conditions:

1. The current analysis of the structural system.
2. The use of tension members having under a dead load a considerable degree of stiffness due to high prestress, and beyond this still sufficient capacity to accommodate the live load.
3. The use of erection methods that ensure that the design assumptions are realized in an economic manner.

Modern cable-stayed bridges present a three-dimensional system where, in general, the deck girder will either be a concrete box or a system of composite structural steel edge girders, composite transverse structure floor beams, and a concrete deck. Supporting parts such as pylons are in compression and inclined cables in tension. Usually the pylons will be in concrete in either type. The important characteristic of such a three-dimensional structure is the full participation of the transverse construction in the work of the main longitudinal structure. This means a considerable increase in the moment of inertia of the construction, which permits a reduction in the depth of the girders and economy in steel. Orthotropic decks for cable-stayed bridges are seldom used today, because in general they are expensive and fabrication- and welding-intensive.

6.13.1 Arrangement of the Stay Cables

According to the various cable arrangements, cable-stayed bridges could be divided into the following four basic systems shown in Figure 6.14:

1. A radial or converging system where all cables are leading to the top of the tower. Structurally, this arrangement is perhaps the best, as by taking all cables to the tower top the maximum inclination to the horizontal is achieved and consequently it needs the smallest amount of steel. The cables carry the maximum component of the dead and live load forces, and the axial component of the deck structure is at a minimum.

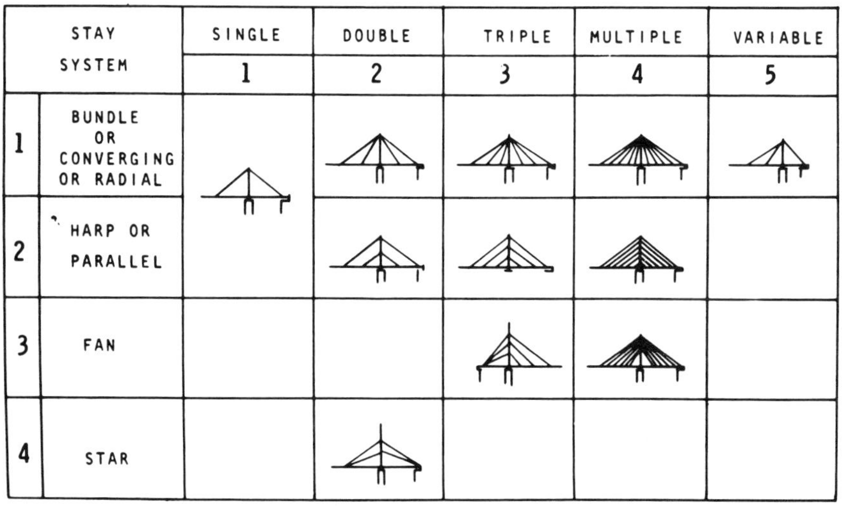

Figure 6.14 Systems of cable arrangement. Reprinted from Troitsky, M. S., *Cable-Stayed Bridges*, 2nd ed., Van Nostrand Reinhold, New York, 1988.

2. Harp or parallel systems have cables connected to the tower at different heights, and placed parallel to each other. This system may be preferred from an aesthetic point of view. However, it causes bending moments in the tower. The harp-shaped cables give an excellent stiffness for the main span if each cable is anchored to a pier. The quantity of steel required for a harp-shaped cable arrangement is slightly larger than for a fan-shaped arrangement.

3. A fan or intermediate system of stay-cable arrangement represents a modification of the harp system. All stays have fixed connections in the tower.

4. The star system is an aesthetically attractive cable arrangement. However, the points of attachment of the cables are not distributed along the main girder.

6.13.2 Positions of the Cables in Space

With respect to the various positions in space that may be adopted for the planes in which the cable stays are disposed, there are two basic arrangements: two-plane systems and single-plane systems (Figure 6.15):

1. The two-vertical-plane system has two alternative layouts: The cable anchorages may be situated on the deck structure, or they may be built inside the main girders. The first layout is the better of the two.

2. Two-inclined-plane system arrangements can be recommended for very long spans where the A-shaped tower has to be very high and needs the lateral stiffness given by the triangle and the frame action.

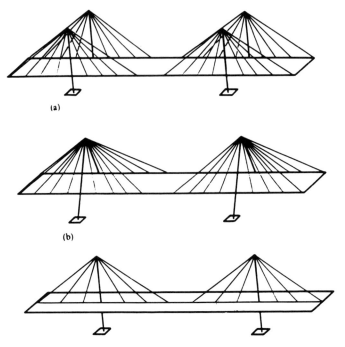

Figure 6.15 Space positions of cables. Reprinted from Troitsky, M. S., *Cable-Stayed Bridges*, 2nd ed., Van Nostrand Reinhold, New York, 1988.

3. The single-plane system along the middle longitudinal axis of the superstructure. In this case, the cables are located in a single vertical strip. This arrangement requires a hollow box main girder with considerable torsional rigidity in order to keep the change of cross-section deformation due to eccentric live load within allowable limits. The single-plane system creates a lane separation as a natural continuation of the highway approaches to the bridge. This is an economical and aesthetically acceptable solution, providing an unobstructed view from the bridge.

6.13.3 Tower Types

Tower types provide various constructions as illustrated in Figure 6.16. Towers may take the form of: (1) trapezoidal portal frames; (2) twin towers; (3) A-frames; or (4) single towers.

There are three different solutions possible regarding the support arrangement of the towers as follows:

1. Towers fixed at the foundation
2. Towers fixed at the superstructure
3. Hinged towers

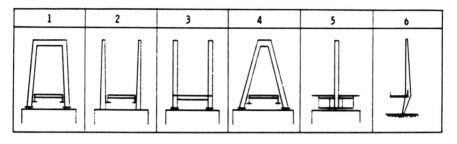

Tower types
1 Portal tower   4 A-frame tower
2 Twin tower   5 Single tower
3 Twin tower   6 Side tower

Figure 6.16 Tower types. Reprinted from Troitsky, M. S., *Cable-Stayed Bridges*, 2nd ed., Van Nostrand Reinhold, New York, 1988.

6.13.4 Deck Types

Orthotropic decks were used for early cable-stayed bridges in Europe (1955–1975). Contemporary cable-stayed bridges seldom use an orthotropic deck. The Sunshine Skyway in Florida (1200-ft central span) uses a concrete trapezoidal single-cell concerte box, the Annacis in Vancouver (1526-ft center span) uses longitudinal edge girders and transverse floor beams and a concrete deck, the Yang Pu bridge in Shangai (1975-ft center span, currently under construction) also uses the longitudinal edge steel box-girder beam, transverse I-plate girder floor beams, and concrete deck. Numerous other examples can be provided.

If an orthotropic deck is used, typical ribs are shown in Figure 6.17. Cross girders are spaced 6–8 ft apart for decks stiffened by flexible ribs, and 15–18 ft apart in the case of decks stiffened by box-type ribs possessing a high degree of torsional rigidity.

For relatively small spans in the 200–300 ft range, it is convenient to use a reinforced concrete deck acting as a composite section with the steel grid formed by the stringers, floor beams, and main girders.

6.13.5 Main Girders and Trusses

Main girders and trusses of the following three types are presently being used for cable-stayed bridges:

1. Steel girders with a solid web may be divided into two types: those constructed with I-girders, and those with one or more enclosed box sections, as shown in Figure 6.18.
2. Plate I girders with a built-up bottom flange comprising a number of cover plates have been used in some bridges.

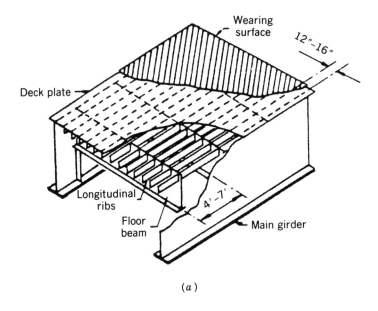

(a)

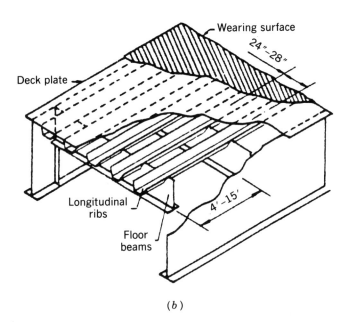

(b)

Figure 6.17 Rib types. Reprinted from Troitsky, M. S., *Orthtropic Bridges Theory and Design*, 2nd ed., The James F. Lincoln Arc Welding Foundation, Cleveland, OH, 1987.

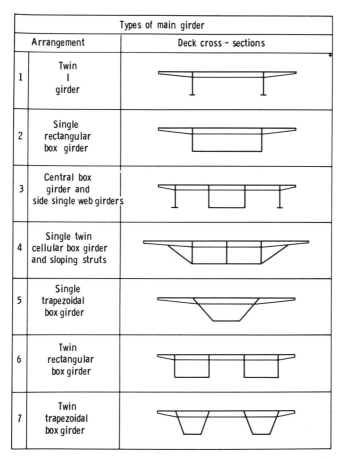

Figure 6.18 Types of main girder. Reprinted from Troitsky, M. S., *Cable-Stayed Bridges*, 2nd ed., Van Nostrand Reinhold, New York, 1988.

3. Box girders have the great advantage of simplicity of fabrication in comparison with plate I girders, and most important, a standard section with only the plate thickness varying can be produced in series, which significantly reduces fabrication costs. Box girders may be rectangular or trapezoidal in form.

Trusses have rarely been used in the construction of cable-stayed bridges during the last decade. They require a great deal of fabrication and maintenance, and protection against corrosion is difficult. However, trusses may be used instead of girders for aerodynamical reasons, and in the case of combined highway and railway traffic, when double-deck structures are usually used. In Figure 6.19 typical bridge cross-sections incorporating trusses are shown.

124 SUPERSTRUCTURE—STEEL BRIDGES

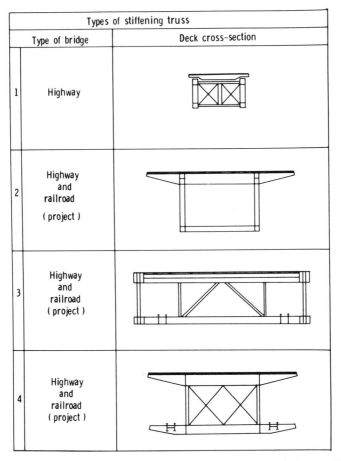

Figure 6.19 Deck supported by stiffening trusses. Reprinted from Troitsky.

6.13.6 Structural Advantages

A deck supported by concrete girders is shown in Figure 6.20. The structural advantages of the cable-stayed system in bridge engineering has resulted in the creation of new types of structures that possess many excellent characteristics and advantages. Outstanding among these are their structural characteristics, efficiency, and wide range of application.

By their structural behavior cable-stayed systems occupy a middle position between girder- and suspension-type bridges. The main structural characteristics of this system are the integral action of the stiffening girders and prestressed or posttensioned inclined cables, which run from the tower tops down to the anchor points at the stiffening girders. Horizontal compressive forces due to the cable action are taken by the girders and no massive anchorages are required. The substructure, therefore, is very economical.

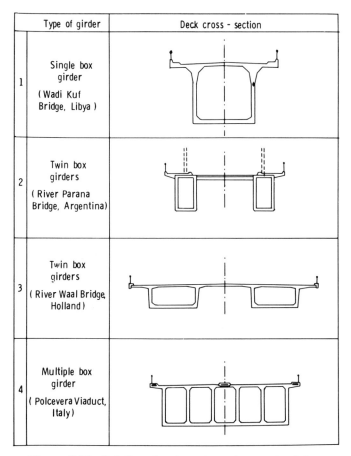

Figure 6.20 Reinforced and prestressed concrete girders.

Introduction of the orthotropic system has resulted in the creation of new types of superstructures that can easily carry the horizontal thrust of stay cables with almost no additional material, even for very long spans.

6.13.7 Comparison of Cable-Stayed and Suspension Bridges

The superiority of the cable-stayed bridge over the suspension bridge may be based on a comparison of their structural characteristics following an analysis as proposed and developed by Gimsing, who found that the relative costs and quantities of steel are substantially smaller for cable-stayed bridges; also deflections are greater at suspension bridges covering the same spans. The outstanding advantage of the cable-stayed bridge is that it does not require as large or as heavy anchorages for the cables as the suspension bridge does.

6.13.8 Composite Cable-Stayed Bridges

In the design of modern cable-stayed bridges, the trend is to use concrete pylons and composite steel girders for the deck system.[8] In this case, the deck consists of a grid of welded steel plate girders with two main girders located at the very outside and tapped by a concrete precast slab. Generally, the advantages of such composite decks, as considered by some authors,[1,2] are as follows:

1. The roadway slab is made of concrete, instead of the usual steel orthotropic deck, to reduce costs.
2. By using precast slabs for the deck, the distribution of compression forced onto the steel girders due to shrinkage and creep is minimized and forming costs are reduced.
3. The roadway spans longitudinally between floor beams, tensile stresses, and compression stresses from the overall system.
4. The deck in the center of the main span is not prestressed, although the overall compression is decreasing there. For crack control, the reinforcement is increased and lapped in cast-in-place joints.
5. Greater resistance against rotation to the torsionally weak bridge deck system is achieved by anchoring the stay cables to the outside main girders and converging them at the tower top, thereby creating a stiff space truss.

The composite-type deck was successfully applied to a number of existing cable-stayed bridges, and this concept was also proposed as a steel alternative for a number of other bridges in the United States.

REFERENCES

1. Heins, C. P. and Firmage, D. A., *Design of Modern Steel Highway Bridges*, Wiley, New York, 1979, pp. 55–70.
2. Protasov, K. G., Teplitskii, A. V., Kramarev, S. Ya, and Nikitin, M. K., *Metal Bridges, Construction and Design*, 2nd ed., Transzhelizdat, Moscow, 1973, pp. 3–87 (in Russian).
3. Tall, L. (ed.), *Structural Steel Design*, 2nd ed., Robert E. Krieger, Malabar, Florida, 1983, pp. 180–243.
4. Kuzmanovic, B and Willems, N., *Steel Design for Structural Engineers*, 2nd ed., Prentice-Hall, Englewood Cliffs, NJ, 1983, pp. 212–263.
5. Merritt, F. S. (ed.), *Structural Steel Designers' Handbook*, McGraw-Hill, New York, 1972, pp. 11-1 to 11-144.
6. Heins, C. P. and Firmage, D. A., op. cit., pp. 89–153.
7. Protasov, K. G., Teplitskii, A. V., Kramarev, S. Ya, Nikitin, M. K., 1973, op. cit., pp. 3–87.

8. Gibshman, E. E., *Design of Metal Bridges*, Transport, Moscow, 1969, pp. 75–91 (in Russian).
9. Tall, L. op. cit., pp. 244–282.
10. Kuzmanovic, B. and Willems, N., op. cit., pp. 433–500.
11. O'Connor, C., *Design of Bridge Superstructures*, Wiley, New York, 1971, pp. 22–115.
12. Merritt, F. S., op. cit., pp. 11-1 to 11-144.
13. Heins, C. P. and Firmage, D. A., op. cit., pp. 70–88.
14. Gibshman, E. E., op. cit., pp. 91–111.
15. Tall, L., op. cit., pp. 446–481.
16. Heins, C. P. and Firmage, D. A., op. cit., pp. 89–153.
17. Gibshman, E. E., *Metal Bridges on Highways*, 3rd ed., Autotransport Literature, Moscow, 1954, pp. 71–72.
18. Gibshman, E. E., Kalmykov, N. Ya, Polivanov, N. I., and Kirilov, V. S., *Bridges and Structures on the Highways*, Autotransizdat, Moscow, 1961, pp. 536–538 (in Russian).
19. Hombly, E. C., *Bridge Deck Behaviour*, Chapman and Hall, London, 1976.
20. Rockey, K. C. and Evans, H. R. (ed.), *The Design of Steel Bridges*, Granada Publishing, London, 1981.
21. Kristek, V., *Theory of Box Girders*, Wiley, New York, 1979.
22. Ilyasevich, S. A., *Metal Box Bridges*, Edition Transport., Moscow, 1970 (in Russian).
23. Heins, C. P. and Firmage, D. A., op. cit., pp. 227–229.
24. Troitsky, M. S., *Orthotropic Bridges, Theory and Design*, 2nd ed., James F. Lincoln Arc Welding Foundation, Cleveland, OH, 1987.
25. Heins, C. P. and Firmage, D. A., op. cit., pp. 159–226.
26. Gibshman, E. E., op. cit., pp. 145–214.
27. O'Connor, C., op. cit., pp. 192–226.
28. Merritt, F. S., op. cit., pp. 12-1 to 12-25.
29. Protasov, K. G., Teplitskii, A. V., Kramarev, S. Ya, and Nikitin, M. K., op. cit., pp. 88–221.
30. Gibshman, E. E., *Metal Bridges on Highways*, 3rd ed., Autotransport Literature, Moscow, 1954, pp. 129–130.
31. Protasov, K. G., Teplitskii, A. V. Kramarev, S. Ya, and Nikitin, M. K., op. cit., pp. 222–271.
32. Gibshman, E. E., op. cit., pp. 291–315.
33. O'Connor, C., op. cit., pp. 488–544.
34. Merritt, F. S., op. cit., pp. 13-1 to 13-45.
35. Steinman, D. B., *A Practical Treatise on Suspension Bridges*, 2nd ed., Wiley, New York, 1953.
36. Pugsley, A., *The Theory of Suspension Bridges*, Edward Arnold, Ltd. London, 1968.
37. Frankland, F. H., *Suspension Bridges of Short Span*, American Steel Construction, 1934.

38. University of Washington Engineering Experiment Station, Aerodynamic stability of suspension bridges, 1930.
39. Tsaplin, S. A., *Suspension Bridges*, Dorizdat, Moscow, 1949 (in Russian).
40. Kachurin, V. K., Brazin, A. V. and Erunov, G. G., *Design of Suspension and Cable-Stayed Bridges*, Transport, Moscow, 1971 (in Russian).
41. Gibshman, E. E., op. cit., pp. 317–368.
42. O'Connor, C., op. cit., pp. 371–454.
43. Merritt, F. S., op. cit., pp. 14–17 to 14–44.
44. Troitsky, M. S., *Cable Stayed Bridges, An Approach to Modern Bridge Design*, 2nd ed., Van Nostrand Reinhold, New York, 1988.
45. Heins, C. P. and Firmage, D. A., op. cit., pp. 421–458.
46. O'Connor, C., op. cit., pp. 455–487.
47. Merritt, F. S., op. cit., 14–52 to 14–66.
48. Podolny, W., Jr. and Scalzi, J. B., *Construction and Design of Cable-Stayed Bridges*, 2nd ed., Wiley, New York, 1986.
49. Gimsing, N. J., *Cable Supported Bridges, Concept and Design*, Wiley, New York, 1983.
50. Walther, R., Houriet, B., Isler, W., and Moïa, P., *Cable Stayed Bridges*, Thomas Telford, London, 1988.
51. Roik, K., Albrecht, G., and Weyer, U., *Cable Stayed Bridges*, Ernst, Vëlag für Architektur und technische Wissenschaften, Berlin, 1986.
52. Ulstrup, C. C., (ed.), Cable stayed bridges, *Proceedings of a Session Sponsored by the Structural Division of the American Society of Civil Engineers*, Nashville, Tennessee, May 9, 1988.

CHAPTER 7

SUPERSTRUCTURE—REINFORCED CONCRETE BRIDGES

7.1 SLAB BRIDGES

A simply supported highway slab bridge consists of a monolithically placed slab which spans the distance between supports without the aid of girders or stringers. The slab bridge is an efficient structure for short spans.[1-5] Because of the weight of the solid slab, it is not economical for long spans. Slab bridges have been used for simple spans up to 35 ft, but many designers find them most economical when they are not more than 20–25 ft. Continuity over the supports increases the economical span length, but at the expense of simplicity in design and field procedures. For simple spans, the span is the distance to the center of supports. Concrete slab bridges are longitudinally reinforced and should also be reinforced transversely to distribute the live loads laterally. The amount should be at least the following percentage of the main reinforcing steel required for positive moment, $100/\sqrt{S}$, but it need not exceed 50% where S = span (ft). The slab should be strengthened at all unsupported edges. In the longitudinal direction, strengthening may consist of a slab section additionally reinforced, a beam integral with and deeper than the slab, or an integral reinforced section of slab and curb.

7.2 DECK-GIRDER BRIDGES

A deck-girder bridge consists of longitudinal main girders with concrete slabs spanning between the girders. The spacing of longitudinal girders or floor beams should be close enough to permit the use of thin slabs so that the dead

130 SUPERSTRUCTURE—REINFORCED CONCRETE BRIDGES

load remains relatively small. Deck-girder bridges have many variations in design and fabrication. Some variations are discussed in the following sections.

7.2.1 Reinforced Concrete T-Beams

This type of bridge, widely used in highway construction, consists of a concrete slab supported on and integral with girders. It is especially economical in the 50- to 80-ft range where falsework is prohibited. Because of traffic conditions or clearance limitations, precast construction of reinforced or prestressed concrete may be used. But adequate bond and shear resistance must be provided at the junction of slab and girder to justify the assumption that they are integral.[6-10]

Deck-girder bridges are simple to design and relatively easy to construct. They are economical to a considerable range of span lengths. Some variations of deck-girder bridges in design and fabrication are:

1. Reinforced-concrete T-beam
   a. Beams and floor cast monolithically
   b. Precast beams and floor cast in place
   c. Precast beams and precast floor sections
2. Prestressed concrete
   a. Prestressed girders and floor cast in place
   b. Precast prestressed girders with reinforced-concrete floor slab cast in place.
   c. Precast prestressed girders with many possible methods of fabricating and placing the floor.

Since the girder is parallel to traffic, the main reinforcing in the slab is perpendicular to traffic. For a simply supported slab, the span should be the distance center to center of supports but need not exceed the clear distance plus thickness of slabs. For slabs continuous over more than two girders the clear distance may be taken as the span.

The ratios of beam depths to spans used in continuous T-beam bridges generally range from 0.065 to 0.075. An economical depth usually results when a small amount of compression reinforcement is required at the interior supports. Girder spacing ranges from about 7 to 9 ft. Usually, a deck slab overhang of about 2 ft 6 in. is economical. When the slab is made integral with the girder, the effective width in design may not exceed the distance center to center of girders, one-fourth the girder span, or 12 times the least thickness of the slab, plus the girder web width. For exterior girders, however, effective overhang width may not exceed half the clear distance to the next girder, one-twelfth the girder span, or six times the slab thickness.

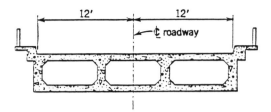

Figure 7.1 Typical cross-section of box girder bridge.

7.3 BOX-GIRDER BRIDGES

A box-girder bridge span consists of longitudinal girders with top and bottom slabs which form hollow or box girders (Fig. 7.1). This type girder is useful for a large range of span lengths.[11-16] Simple spans as small as 40 ft have been used, but in general the box girder spans of reinforced concrete are more economical in the range of 60 to 100 ft and usually are designed as continuous structures over the piers. Box girders of prestressed concrete in the design of which the advantages of continuity are utilized, have been built with span lengths of approximately 300 ft.

An important feature of the box girder is its resistance to torsional loadings. It can be built to curve horizontally to conform with road alignment, follow vertical curves, and tilt sideways to provide varying superelevation. The piers do not need to be complete bents but may be individual columns staggered to free traffic lanes below an overpass structure. Box girders are particularly adaptable as continuous structures. The top or bottom slab may be thickened depending on whether positive or negative bending movements are present. The girder webs may be thickened in proportion to the shear.

For sites where structural depth is not severely limited, box girder and T-beams have been about equal in price in the 80-ft span range. For shorter spans, T-beams usually are cheaper, and for longer spans, box girders. While these cost relations hold in general, box girders have, in some instances, been economical for spans as short as 50 ft when structural depth was restricted.

7.3.1 Box Girder Design

Structural analysis is usually based on two typical segments, interior and exterior girders (Fig. 7.2). The requirements listed in Standard Specifications for Highway Bridges (AASHTO) are based on live-load distributions for individual girders and so design usually is based on the assumption that a box-girder bridge is composed of separate girders.

The effective width of slab as the top of an interior girder may be taken as the smallest of the distance center to center of girders, one-fourth the girder

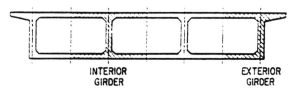

Figure 7.2 Typical design sections—cross hatched for box-girder bridges.

span, and 12 times the least thickness of slab, plus girder web width. Effective overhang width for an exterior girder may be taken as the smallest of half the clear distance to the next girder, one-twelfth the girder span, and six times the least thickness of the slab. The commonly used depth-to-span ratio for continuous spans is 0.065.

Box-girder bridges are pleasing in appearance because of the long continuous lines which may be obtained. For spans of varying length, the depth may be held constant, the bottom surface can be made simple and clean in appearance, and continuous curves rather than short straight chords are possible.

7.4 PRESTRESSED CONCRETE SEGMENTAL BRIDGES

Prestressed concrete segmental bridge construction has evolved from combining the concepts of prestressing, box girder design, and the cantilever method of bridge construction.[17-19] It arose from the need for a bridge erection method that dispensed with the use of conventional falsework.

Contemporary prestressed, box girder, segmental bridges began in Western Europe in 1950, with the first application in Canada in 1964, and in the United States in 1973. Before 1981 in the United States more than 80 segmental bridges were completed. Prestressed concrete segmental bridges may be identified as precast or cast in place and categorized by method of construction as balanced cantilever, span-by-span, progressive placement, or incremental launching. This type of bridge is adaptable to almost any conceivable site condition.

7.4.1 Cast-In-Place Balanced Cantilever

Cantilever construction, whether cast in place or precast, may proceed from permanent piers and the structures are self-supporting at all stages (Fig. 7.3). In cast-in-place construction the framework is supported by a movable form carrier (Fig. 7.3). The form traveler moves forward on rails attached to the deck of the completed structure and is anchored to the deck at the rear. With the form traveler in place, a new segment is formed, cast, and stressed to the previously constructed segment.

7.4 PRESTRESSED CONCRETE SEGMENTAL BRIDGES

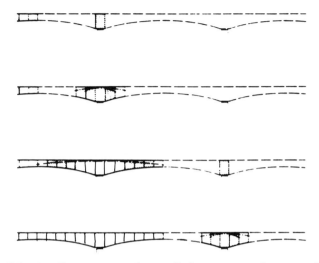

Figure 7.3 Cantilever construction applied to prestressed concrete bridges.

7.4.2 Precast Balanced Cantilever

The erection option available can be adopted to almost all construction sites:

a. *Crane Placing.* Truck or crawled. Cranes are used on land where feasible; floating cranes may be used for a bridge over navigable water.

b. *Beams and Winch Method.* If access by land or water is available under the bridge deck, segments may be lifted into place by hoists secured atop the previously placed segments.

c. *Launching Gantries.* There are two families of launching gantries. The first family is shown in Figure 7.4. The launching gantry is slightly more than the typical span length, and the gantry's rear support reaction is applied near the far end of the last completed cantilever. All segments are brought onto the finished deck and placed by the launching gantry in the balanced cantilever. In the second family the launching gantry has a length approximately twice the typical span (Fig. 7.5).

7.4.3 Span-By-Span Construction

For long viaduct structure, which relatively shorter spans, span-by-span methodology was developed using a form traveler with construction joints or hinges located at the points of contraflexure. The form traveler may be supported on the piers, or from the edge of the previously completed construction. In the initial form, the span-by-span method is a cast-in-place technique.

134 SUPERSTRUCTURE—REINFORCED CONCRETE BRIDGES

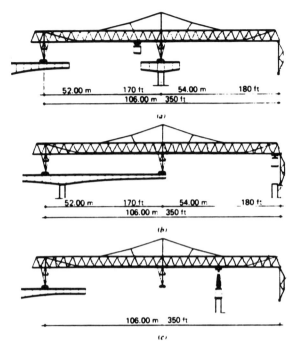

Figure 7.4 First family of launching gantries.

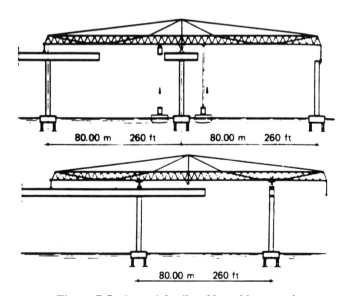

Figure 7.5 Second family of launching gantries.

7.4.4 Progressive Placement Construction

Progressive placement is similar to the span-by-span method in that construction starts at one end. In progressive placement, the precast segments are placed from one end of the structure of the other on successive cantilevers. At present, this method appears practicable and economical in spans from 100 to 300 ft (30 to 90 m). The erection procedure is illustrated in Figure 7.6.

7.4.5 Incremental Launching or Push-Out Construction

Segments of the bridge superstructure are cast in place in lengths of 50-100 ft (10-30 m) in stationary forms located behind the abutment(s). Each unit is cast directly against the previous unit. After sufficient concrete strength is reached, the new unit is posttensioned to the previous one. The assembly of units is pushed forward in a stepwise manner to permit casting of the succeeding segments.

7.4.6 Range of Application of Bridge Type by Span Lengths Considering Segmental Construction

Span (ft)	Bridge Types
0-150	I-type pretensioned girder
100-300	Cast-in-place posttensioned box girder
100-300	Precast-balanced cantilever segmental, constant depth
200-600	Precast-balanced cantilever segmental, variable depth
200-1000	Cast-in-place cantilever segmental
800-1500	Cable-stay with balanced cantilever segmental

7.5 REINFORCED CONCRETE TRUSSES

Reinforced concrete truss-type bridges (Fig. 7.8), having spans more than 100 ft (30 m) are very rare. Bridges having moderate spans of 65 ft (20 m) are more numerous.[20,21] Reinforced concrete trusses are generally of two types—those having only verticals and no diagonals, the so-called Vierendel trusses (Fig. 7.7). At spans of more than 100-150 ft (30-35 m) trusses have the top chord of curvilinear configuration and the ratio of the height to span is $h/l = 1/7-1/8$ (Fig. 7.8a, b). Trusses having parallel chords (Fig. 7.8c) by comparison with the girders provide a reduction in volume approximately up to 40-60%. Such systems for spans 65-80 ft (20-25 m) have ratio $h/l = 1/8 - 1/10$. The length of the panel at trusses having parallel chords is h to $0.9h$. Some reinforced concrete trusses are similar to the metal trusses (Fig. 7.8). Bridges having middle-size spans have parallel chords (Fig. 7.8a-c) and larger spans have an upper chord of parabolic shape (Fig. 7.8c). The slope of the diagonals is similar as in metal bridges, namely, 40-50 degrees.

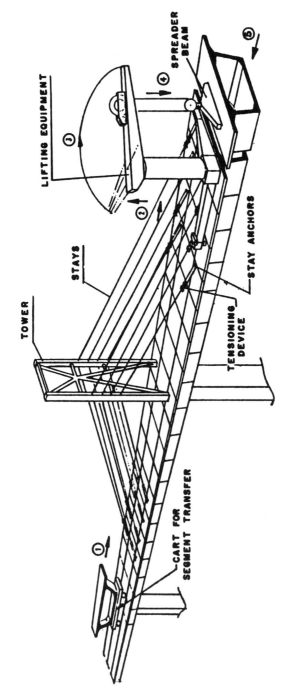

Figure 7.6 Progressive placement construction.

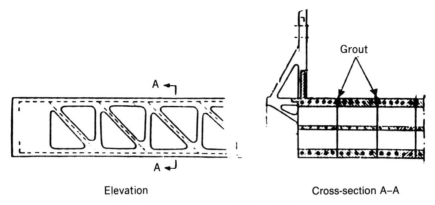

Figure 7.7 Schemes of Vierendel trusses.

7.6 FRAME BRIDGES

The rigid frame bridge is a structure in which the columns are made so stiff that the connecting girders are fixed ends.[22-25] This makes possible a reduction in depth at the center of the span, thereby reducing the dead load where it is most harmful. It further makes possible the design of a shallower girder, or slab, to satisfy both strength and stiffness requirements. This system proves its advantages both from an economic and architectural point of view and has now largely superseded the arch for separating grades at intersections of highways and railways.

The bases of the columns are usually approximately hinged; the section here is relatively thin and not capable of supporting a large resisting moment. The frame column often acts as an abutment wall; hence the outward reaction from the frame opposes the inward earth trust. In some designs, however, the abutment wall and frame back leg are separate. The layout may be for slab design or for separate ribs with a transverse floor system. For highway overcrossing, a single span is preferable, but multiple spans are needed if the crossing is very wide.

The structural analysis of rigid frames involves no special principles differing from those in ordinary frames; the fixed-end moments are greater because the center section is less stiff. The columns are subject to high moments.

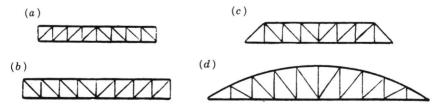

Figure 7.8 Schemes of reinforced concrete trusses.

7.7 ARCHES

7.7.1 Introduction

Reinforced-concrete arches are designed to resist tensile stresses if necessary.[26] Since concrete is relatively weak in tension and strong in compression, its use in arch structures is especially logical. Although much longer spans have been built, the average reinforced concrete arch has a span length between 80 and 200 ft. Concrete arches are pleasing in appearance and because of this are often chosen even if a cost premium is involved.

7.7.2 Structural System

The stress distribution in an arch depends on the number of hinges.[27] A fixed arch (Fig. 7.9a), is statically indeterminate to the third degree; a single-hinged arch (Fig. 7.9b) is statically indeterminate to the second degree; a two-hinged arch (Fig. 7.9c) is statically indeterminate to the first degree; and the three-hinged arch is statically determinate. In reinforced concrete, the fixed arch is the most common. The three- and two-hinged types are more common than the single-hinged type.

The choice of number of hinges depends on the particular conditions of design. Fixed arches require foundations which resist vertical, horizontal, and rotational movements. The stresses in a two-hinged arch are not sensitive to rotational or to differential settlement of the foundation.

Various parts of the arch bridge are illustrated in Figure 7.10. Figure 7.10(a)

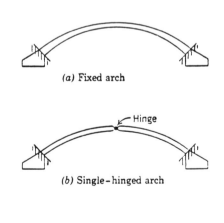
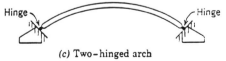

Figure 7.9 Hinge arrangement in arches.

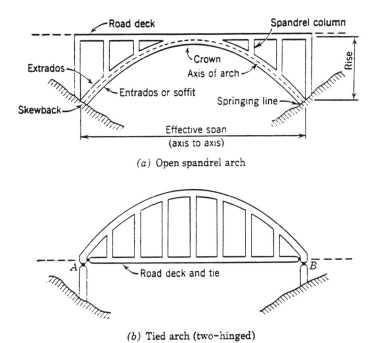

Figure 7.10 Arch bridge types.

is the open spandrel type, which means that the roadway is supported above the arch by columns and girders.[28] If the spandrel is closed by a retaining wall and the interior is filled with a light material so that the roadway is placed directly on the fill, the structure is known as a spandrel-filled arch bridge. The tied arch (Fig. 7.10b) is used when design conditions indicate that the arch be placed above the roadway. A sensible member is suspended at the roadway elevation, which provides the horizontal force required for arch action.

7.7.3 Abutment of the Arch

The important dimensions of the axial curve of an arch are the span and the rise. These are generally controlled by local conditions such as channel width, vertical clearance, and high water conditions. If local conditions do not dictate these dimensions, various economic factors are considered. An arch with a large rise-to-span ratio develops relatively less horizontal thrust, but the structure is relatively heavy. A small rise-to-span ratio produces large horizontal thrusts, and the increase in foundation costs must be included. The stresses due to temperature, shrinkage, and elastic shortening are relatively large for flat arches and require extra material. Most arch bridges have a rise-to-span ratio between 0.15 and 0.30.

140 SUPERSTRUCTURE—REINFORCED CONCRETE BRIDGES

7.7.4 Geometry of the Arch

Once the span and rise are chosen, three points on the arch axis are located. The shape of the arch at the other points is influenced by the location of the thrust line and by appearance. An arch in the form of a smooth curve is considered pleasing in appearance. Based on estimated dead load plus live load, the thrust polygon which passes through the three loaded points is determined first. Then the arch axis is shaped to conform as closely as practical to the thrust line.

Several combinations of smooth curves are used to approximate the thrust-line polygon. The curve can be circular, elliptical, parabolic, or multicentered. The radial thickness of a fixed arch should increase from the crown to the springing line. The radial thickness of the ring is often made equal to the thickness at the crown multiplied by the secant of the slope angle. This suggests a ratio of springing line to crown thickness of from 1.5 to 2.2.

As in any statically indeterminate structure, the size of the arch rib must be assumed for preliminary design and adjustments made before a final design is determined.

7.7.5 Robert Maillart's Bridges

A famous Swiss bridge designer, Robert Maillart, introduced original bridge systems consisting of arch disks, which are thrust structures having curvilinear bottom chords and horizontal top configurations. In the system consisting of arch disks in which the roadway and arches are connected monolithically, the idea of connecting all parts of the bridge-arches, columns, and roadway for common work resulted in architecturally beautiful bridge structures.[29,30]

An arch disk consists of a longitudinal wall, connected at the top by the roadway slab and at the bottom by the plate of the vault. Depending on the bridge width, there may be two or more of Maillart's arch disk system longitudinal walls, which usually consist of a three-hinged arch.

Another of Maillart's great achievements is in modern architectural shaping of concrete bridges. Some typical examples are:

1. Reinforced concrete highway bridge, Salginatobel, Switzerland, has a three-hinged arch, span of 295 ft (90.4 m), width of 11.5 ft (3.50 m) (Figs. 7.11 and 7.12) and was built in 1929–1930.
2. Reinforced concrete Schwandbach bridge, built in 1933, having elliptical curvature in the plan (Figs. 7.13 and 7.14)
3. Pedestrian bridge Töss, built 1934 (Fig. 7.15).
4. Bridge Champel-Vessy, built 1936–1937.

7.7.6 Concrete Suspension Bridges

Concrete suspension bridges have a reinforced or a prestressed concrete deck and pylons.[31] The deck generally consists of a reinforced or prestressed con-

Figure 7.11 General view of Salginatobel bridge.

crete slab supported by transverse steel beams whose ends are connected to the suspenders. The reinforced or prestressed concrete slab may also be supported by reinforced or prestressed concrete transverse beams and each end of these beams is connected to the suspenders. The advantage of a concrete deck in suspension bridges is its ability to take wind pressure; therefore, it is not necessary to provide special wind bracing.

Reinforced or prestressed concrete pylons have either hinged or fixed supports. Pylons having fixed supports are essentially vertical cantilevers and offer relatively small resistance to cable movement spanwise. Concrete pylons have found application in suspension bridges with relatively small spans up to 850 ft.

Reinforced concrete pylons may be built using flexible bars or a rigid steel skeleton. The advantage of a steel skeleton is that it does not require falsework for concreting. It should be noted that pylons having fixed supports are mostly applied in concrete suspension bridges.

7.8 CONCRETE CABLE-STAYED BRIDGES

In the last decade, a number of cable-stayed bridges having a reinforced or prestressed concrete deck system have been erected.[32-38] These structures possess a high degree of rigidity and a relatively small deflection, and their damp-

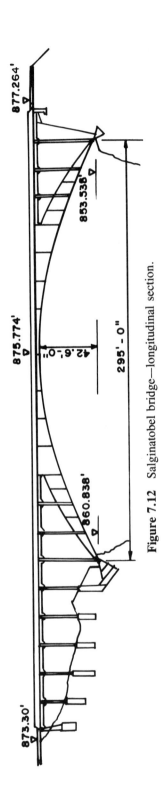

Figure 7.12 Salginatobel bridge—longitudinal section.

7.8 CONCRETE CABLE-STAYED BRIDGES

Figure 7.13 Schwandbach bridge.

ing effect is such that there are relatively small vibrations. Most often this system consists of a stiffening girder separated into two cantilevers which are connected by a middle single-span girder. This system is very convenient for a free cantilevering construction.

Multispan structures could be very economical; usually these spans are equal, so the dead load is compensated. The main advantages can be summarized as follows:

1. The horizontal components of the inclined cable forces, causing compression combined with the bending, very much favor a deck system design using monolithic or precast concrete.
2. The depth of the main girder is very shallow, like that of a tied arch bridge with a suspended deck.
3. The amount of steel used for the cables is comparatively small. An optimum solution can be achieved by correctly choosing the height of the tower.
4. The erection of the cables, as well as the reinforced concrete deck, is comparatively easy. Also, construction with the free cantilevering system is very suitable. Owing to the small amount of steel and the ease of erection, this system can be highly recommended with regard to the cost.
5. The deflections are small and, therefore, this system is applicable for railroad bridges.

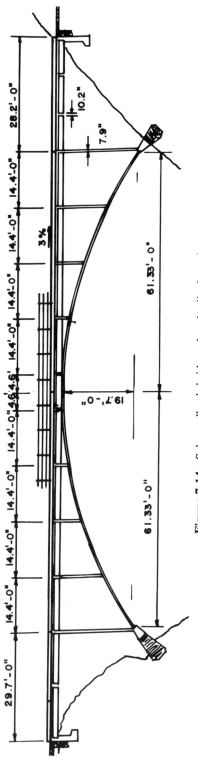

Figure 7.14 Schwandbach bridge—longitudinal section.

Figure 7.15 Bridge Töss.

6. Such outstanding structures as, for instance, the Maracaibo bridge in Venezuela, indicate that this new bridge system possesses many excellent characteristics.

7.9 PRESTRESSED CONCRETE BRIDGES

7.9.1 Prestressing Methods

There are two methods of tensioning elements for prestressed concrete: pretensioned and posttensioned.[39-44] The prestressing tendons in a given member will be all one kind or another, or a combination of the two, depending upon conditions. The term pretensioned means that the tendons are tensioned to their full load before the concrete is placed. They are held under tension by anchors beyond the ends of the prestressed concrete number. After the concrete has been placed and allowed to cure to sufficient strength, the load in the tendons is transferred from the external anchors into the newly poured member, thus prestressing it.

The term posttensioned means that the tendons are tensioned after the concrete has been placed and allowed to cure. Frequently, the tendon is placed inside a flexible metal duct, and the entire assembly is poured around it. After the concrete has cured, the tendon is tensioned and held under load by anchor

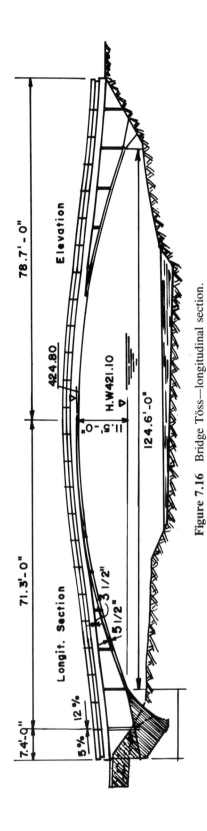

Figure 7.16 Bridge Töss—longitudinal section.

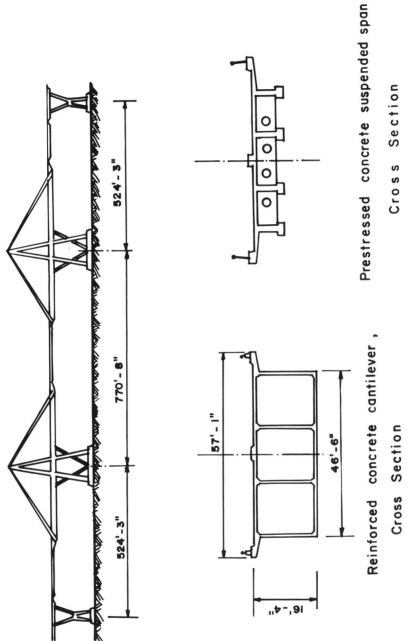

Figure 7.17 Reinforced and prestressed concrete girders.

fittings at its ends. Bond between the tendon and the concrete member is achieved by pumping the metal duct full of grout.

7.9.2 Economy

The savings in material by using prestressed concrete for bridges, according to Lin, is as follows: The weight of prestressing steel per square foot of bridge is about 1/5 that of the reinforcing steel in a reinforced concrete bridge. Prestressing allows considerably better utilization of concrete than conventional reinforcement. It results in an overall dead-load reduction, which makes long spans possible with concrete, and sometimes competitive in cost with those of steel. Prestressed concrete, however, requires higher-quality concrete and steel, and greater refinement and control in fabrication than does reinforcing concrete. Depending on the methods and sequence of fabrication, prestressed concrete may be precast-pretensioned, precast-posttensioned, cast-in-place postensioned, or composite.

7.9.3 Precast-Beam Bridges

These widely used structures consist of precast-prestressed I-beams, T-beams, and box girders, which may be either pretensioned or posttensioned.[45-48] Precast I-beams may be built with cast-in-place decks. With a precast, prestressed T-beam as with an I-beam, the flange must be connected with cast-in-place concrete. Precast, prestressed box sections may be placed side by side to form a bridge span. If necessary, they may be posttensioned transversely.

Precast, prestressed beams are used mainly for spans up to about 120 ft at locations where erection of falsework is impossible or not desirable. Such beams are economical for mass fabrication. For longer spans it is necessary to provide heavy equipment for erection and/or transporting purposes.

7.9.4 Cast-In-Place Prestressed Concrete

For this type of prestressed concrete, typical cross-sections are generally similar to those used for conventionally reinforced sections, but prestressing permits thinner-walled designs. For a fully cast-in-place single-span bridge, post-tensioning differs only quantitatively from that for precast elements.

7.9.5 The Cantilever System

This system was developed for continuous prestressed concrete sections. Starting from each main pier, segments of superstructure are added symmetrically on opposite sides of the piers. Each segment is clamped to those already erected with tensioned rods or cables. The segments may be precast or cast in place. The cantilever arms are connected when they reach the center, usually by a cast-in-place closure concrete placement. Among the advantages of cantilever construction are clear stress patterns, because there are only negative bending moments in the cantilever arms and only positive moments in the suspended spans.

REFERENCES

1. Polivanov, N. I., *Reinforced Concrete Bridge on Highways*, Avtotransizdat, Moscow, 1956, pp. 73–77 (in Russian).
2. American Concrete Institute, Second International Symposium, Concrete Bridge Design, Chicago, April 2, 1969, pp. 75–182.
3. Merritt, F. S. (ed.), *Standard Handbook for Civil Engineers*, 2nd ed., McGraw-Hill, New York, 1976, pp. 17-36 to 17-43.
4. Hambly, E. C., *Bridge Deck Behaviour*, Wiley, New York, 1976, pp. 46–68.
5. Hein, C. P. and Lawrie, R. A., *Design of Modern Concrete Highway Bridges*, Wiley, New York, 1984, pp. 52–131.
6. Polivanov, N. I., op. cit., pp. 54–72.
7. Vlasov, G. M., Geronimus, V. B., Povalyaev, E. V., Ustinov, V. P., and Jacobson, K. K., *Design of Reinforced Concrete Bridges*, Edition Transport, Moscow, 1970, pp. 111–143 (in Russian).
8. Merritt, F. S., op. cit., pp. 17-43 to 17-47.
9. Hambly, E. C., op. cit., pp. 70–86.
10. Heins, C. P. and Lawrie, R. A., op. cit., 132–202.
11. American Concrete Institute, op. cit., pp. 272–379.
12. Merritt, F. S., op. cit., pp. 17-47 to 17-49.
13. Hambly, E. C., op. cit., pp.87–115.
14. Degenkolb, O. H., *American Concrete Box Girder Bridges*, American Concrete Institute, Detroit, Michigan, 1977.
15. Schlaich J. and Scheff, H., *Concrete Box-Girder Bridge*, IABSE, International Association for Bridge and Structural Engineering, Zürich, 1982.
16. Heins, C. P. and Lawrie, R. A., op. cit., pp. 203–282.
17. Prestressed Concrete Institute, *Precast Segmental Box Girder Bridge Manual*, Chicago, Illinois, 1978.
18. Heins, C. P., and Lawrie, R. A., op. cit., pp. 421–470.
19. Podolny, W. Jr. and Muller, J. M., *Construction and Design of Prestressed Concrete Segmental Bridges*, Wiley, New York, 1982.
20. Polivanov, N. I., op. cit., pp. 148–154.
21. Evgrafov, G. K. and Bugdanov, N. N., *Design of Bridges*, Transport, Moscow, 1956, pp. 315–333 (in Russian).
22. Taylor, F. W., Thompson, S. E., and Smulski, E., *Reinforced-Concrete Bridges*, Wiley, New York, 1950, pp. 268–325.
23. Polivanov, N. I., op. cit., pp. 384–431.
24. Abbett, R. W., *American Engineering Practice*, vol. III, Wiley, New York, 1957, pp. 29-80 to 29-97.

25. Vlasov, G. M., Geronimus, V. B., Povalyaev, E. V., Ustinov, V. P., and Jacobson, K. K., op. cit., pp. 169–186.
26. McCullough, C. B. and Thayer, E. S., *Elastic Arch Bridges*, Wiley, New York, 1948.
27. Polivanov, N. I., op. cit., pp. 434–587.
28. Vlasov, G. M., Geronimus, V. B., Povalyaev, E. V., Ustinov, V. P., and Jacobson, K. K., op. cit., pp. 181–205.
29. Max Bill, *Robert Maillart*, 2nd ed., Girsberger, Zürich, 1955.
30. Billington, D. P., *Robert Maillart's Bridges*, Princeton University Press, Princeton, NJ, 1979.
31. Tsaplin, S. A., *Suspension Bridges*, Dorizdat, Moscow, 1949, pp. 162–163; 186–188 (in Russian).
32. Simons, H., Wind, H., Moser, W. H., *The Bridge Spanning Lake Maracaibo in Venezuela*, Bauverlag GmbH, Wiesbaden, Berlin, 1963.
33. Kireenko, B. I., *Cable-Stayed Bridges*, "Budiveljnik," Kiev, 1967, pp. 25–27 and 81–92 (in Russian).
34. Podolny, W. Jr. and Scalzi, J. B., *Construction and Design of Cable-Stayed Bridges*, 2nd ed., Wiley, New York, 1986, pp. 12, 15–18, 28, 33, 34, 50, 107, 119–124, 128, 150.
35. Troitsky, M. S., *Cable-Stayed Bridges, An Approach to Modern Bridge Design*, 2nd ed., Van Nostrand Reinhold, New York, 1988, pp. 114–146.
36. Abbett, R. W., op. cit., pp. 29–17.
37. Merritt, F. S., op. cit., pp. 17–49.
38. Troitsky, M. S., op. cit., pp. 147–154.
39. Polivanov, N. I., op. cit., pp. 310–350.
40. Lin, T. Y., *Design of Prestressed Concrete Structures*, 2nd ed., Wiley, New York, 1963.
41. American Concrete Institute, op. cit., pp. 663–741.
42. Prestressed Concrete Institute, *Post-Tensioned Box Girder Bridges, Design and Construction*, Chicago, 1971.
43. Podolny, W., Jr., and Muller, J. M., *Construction and Design of Prestressed Concrete Segmental Bridges*, Wiley, New York, 1982.
44. Heins, C. P. and Lawrie, R. A., op. cit., pp. 283–361.
45. American Concrete Institute, op. cit., pp. 752–805.
46. Heins, C. P. and Lawrie, R. A., op. cit., pp. 362–420.
47. Prestressed Concrete Institute, *Precast Prestressed Concrete Short Span Bridges, Spans to 100 Feet*, 2nd ed., Chicago, IL, 1988.
48. Prestressed Concrete Institute, *Design Supplement to: Precast Prestressed Concrete Short Span Bridges, Spans to 100 Feet*, 2nd ed., Chicago, IL, 1988.

CHAPTER 8

SUBSTRUCTURE—PIERS

The principal parts of a bridge are the substructure and the superstructure. Substructure generally includes piers and abutments, which may be of several different types.

8.1 PIERS

A bridge pier is a structure, usually of concrete, which is used to transmit load from the bridge superstructure to the foundation.[1-3] A pier should disturb the natural flow of the water as little as possible. There are many ways in which a particular bridge pier may be built. Every pier should satisfy the following basic requirements: function, architecture, and minimum cost.

Considering the shape of the bridge pier, some of the common parts are as follows[4] (Fig. 8.1):

1. *Bridge pad.* A block of concrete resting on top of the pier to support a base plate and bearings.
2. *Bridge seat.* The top of the pier, to support a pad.
3. *Coping.* The top course of the pier, usually projecting beyond the lower courses. The coping course serves to protect the pier from the weather.
4. *Belting course.* The main function of the belting course is to strengthen the coping offset, but it also improves the appearance of the pier.
5. *Footing courses.* Those courses at or near the bottom of the pier. The function of the footing course is to distribute the load over a larger area

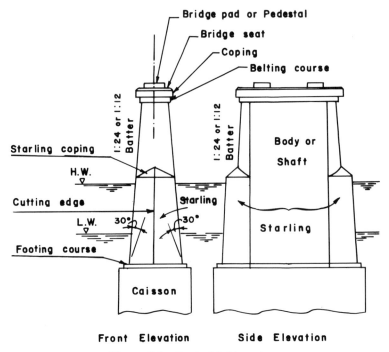

Figure 8.1 Typical bridge pier.

than the base of the shaft of the pier. Unless reinforced, the slope of the footing should not be over 30° with the vertical.

6. *Shaft.* The main part of the pier.
7. *Starling.* That part of the pier below high water, the horizontal section of which lies outside of the largest rectangle that can be formed on the two sides of the pier. The function of the starling is to pass the water with the least possible disturbance, for then there will be the least pressure against the pier due to current, ice, and drift, less danger to navigation from eddies, and less danger from underscouring.
8. *Starling coping.* The offset course at about high water which forms the top course of the starling.
9. *Batter.* The slope of the sides and ends of the pier. Usually a batter slope is 1:24 or 1:12. The smaller ratio is more commonly used for high piers and the greater for low piers.

8.2 PIERS WITH ICE-BREAKING CUTWATER

Sometimes a pier should be designed not merely as a support for the vertical load of the bridge but as a rigid streamlined mass that can withstand ice pressure. In such cases the cutwater is sloped as an inclined wedge (Fig. 8.2). The ice then tends to slide up the starling until one edge is out of the water and its

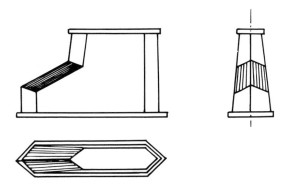

Figure 8.2 Pier with ice-breaking cutwater.

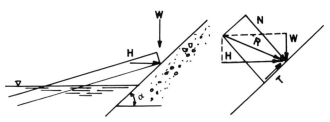

Figure 8.3 Cutwater.

weight is sufficient to break the cake over the "knife edge" (Fig. 8.3). The current then carries the pieces past on both sides of the pier. Usually the cutting edge of a starling is reinforced with a heavy steel angle or with old rails, or it may be made of rounded stones or concrete. What is desired is sliding and breaking, not cutting of the ice.

The position and slope of the starling are determined in accordance with probable flood levels and ice conditions at the particular site concerned. There may be places where ice (or log) jams are likely to occur in spite of the starlings and piers. This is a difficult problem to handle. Stone-filled cribs or concrete blocks with starlings may be located upstream from the piers in the hope that they will break up the ice before it reaches the bridge, or that they may cause the ice jam to occur between them rather than at the piers.

Instead of rails the steel-plate protection from the starlings may be used. Steel plates 1/4 in. thick extend from the river bottom to above high water. They are anchored to the pier by Z-shaped straps 18 in. long, spaced 18 in. apart, and staggered on the nose.

8.3 MATERIALS AND CONSTRUCTION

Before about 1880 it was the universal rule to build piers entirely of stone-masonry. However, at the present time most piers are built entirely of concrete or concrete and stone facing. Piers are sometimes built using a cluster of steel piles. For small and temporary structures, timber may be used.

154 SUBSTRUCTURE—PIERS

The advantage of concrete over stonemasonry is that it costs less. Although its compressive strength is somewhat less than that of first-class stonemasonry, because of its monolithic character, most engineers agree that it is the more suitable material, except possibly for the facing of the pier.

There are some advantages to using a facing of stonemasonry; among these are the more attractive appearance of the pier and the elimination of surface cracks. These surface cracks, almost always present in plain concrete piers, are due to the expansion and contraction, caused by temperature changes of the outer layer of concrete.

Where the all-concrete pier is used, it is advisable to place reinforcing rods near the surface. This reinforcement will prevent the occurrence of, or at least

TABLE 8.1 General Shapes of Piers for Small Bridges

(a) Solid shaft, curved end.
(b) Simple columns and portals.
(c) Solid shaft and starling.
(d) I-type with triangular end.
(e) Solid shaft, triangular end.
(f) Columns and portals.
(g) Slender solid shape with rounded ends.
(h) Rigid frame.

decrease the size of, the cracks noted above, and will also add an element of safety by taking any tensile stresses in the concrete. Reinforcement in horizontal planes under the coping and above the bottom of the footing serves to carry the loads more uniformly into the pier and foundation.

8.4 TYPES OF PIERS

There are many ways in which the type of superstructure to be built will either control or suggest the type of pier construction to be used, at least for the portion above water or ground.[7,8] There may be situations in which the type

TABLE 8.2 General Shapes of Piers for Small Bridges

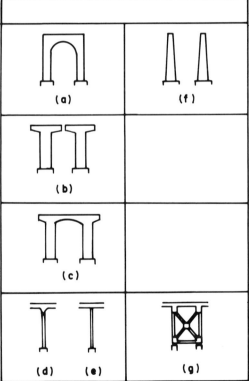

(a) Two columns and portal.
(b) Cantilevered piers for double bridge.
(c) Cantilever and portal combined.
(d) Steel column rigid frame.
(e) Steel column rocker bent.
(f) Two separate columns.
(g) Braced tower.

156 SUBSTRUCTURE—PIERS

of pier to be used is practically dictated by soil conditions and construction procedures.

A pier should have sufficient area at its top to receive the bearings. It should appear strong rather than weak or flimsy. It should be capable of supporting the lateral and longitudinal loads as well as the vertical ones. It should be practical and durable, requiring a minimum of maintenance. And it should be attractive. In general, a pier is most satisfactory if it is simple, neat, and obviously appropriate.

In general:

1. A bridge with two trusses or girders is well adopted to support a two-column type of concrete pier.

TABLE 8.3 General Shapes of Piers for Large Bridges

(a) Vertical shafts with heavy portal and base.
(b) Solid shaft with vertical stepped sections.
(c) Circular shafts and bottom rib.
(d) Shafts battered on three outer faces.
(e) Type with solid ribs battered on two sides.
(f) Triple column for four trusses or girders.

2. A wide bridge with three or four deck trusses or girders may best be held by a corresponding number of pier shafts.
3. A superstructure of closely spaced concrete or steel girders may be placed upon a solid T-pier, a two-column and portal type, or a series of columns.
4. Where mass is required, a solid shaft may be used regardless of the number and position of the bearings. Also, much can be learned from a study of existing types.

The general shapes of some piers for small and large bridges are shown in Tables 8.1–8.4.

TABLE 8.4 General Shapes of Piers for Large Bridges

(a) Shafts stepped with vertical sections fluting.
(b) Cantilevered top shaft.
(c) Triple columns for three trusses or girders.
(d) Shafts partially stepped on three sides.
(e) End view of flared shaft of column.

REFERENCES

1. Abbott, R. W., *American Civil Engineering Practice*, vol. III, Wiley, New York, 1957, pp. 26-22 to 26-30.
2. Kolokolov, N. M., Kopac, L. N., and Fainshtein, I. S., *Artificial Structures*, 3rd ed., Transport, Moscow, 1988, pp. 212-235 (in Russian).
3. Platonov, E. V., *Supports of the Bridges*, 2nd ed., Transzheldorizdat, Moscow, 1946, pp. 112-129 (in Russian).
4. Jacoby, H. S. and Davis, R. P., *Foundations of Bridges and Buildings*, 3rd ed., McGraw-Hill, New York, 1941, pp. 432-466.
5. Jacoby, H. S., and Davis, R. P., op. cit., p. 452.
6. Platonov, E. X., op. cit., pp. 518-562.
7. Taylor, F. W., Thompson, S. E., and Smulski, E., Concrete Plain and Reinforced, Wiley, New York, 1950, pp. 424-434.
8. Gibshman, E. E., Kalmykov, N. J., Polivanov, N. I., and Kirillov, V. S., *Bridges and Structures on the Highways*, Avtotransizdat, Moscow, 1961, pp. 435-441 (in Russian).

CHAPTER 9

SUBSTRUCTURE—ABUTMENTS

9.1 INTRODUCTION

Bridge abutments are structures at the two ends of a bridge used for the double purpose of transferring the loads from the bridge superstructure to the foundation bed and giving lateral support to the approach embankment. In the case of a river crossing, an abutment may have a third function—namely, to protect the embankment against scour of the stream.[1,2]

Determining the locations of the abutments of a bridge is only one of the problems incident to the general planning of the structure. The abutment serves both as a pier and as a retaining wall and in the more usual type consists of a breast wall and wing walls (Fig. 9.1).

9.2 TYPES OF ABUTMENTS

The common types of abutments are listed below.[3,4]

9.2.1 Wing-Type Abutment

In the wing-wall type the wings may be at any angle with the breast wall. In many locations angles of 30 and 45 degrees are widely used (Fig. 9.2). The wing-wall abutment is used to protect the embankment against washing out under high water as well as to furnish better hydraulic conditions by proving a gradual change in the cross-section of the channel. This type may be easily adopted for use with a skewed structure. In general, such an abutment is heavy and is most suitable on firm soils. The wings are designed as retaining walls.

160 SUBSTRUCTURE—ABUTMENTS

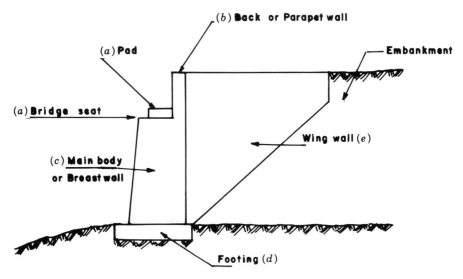

Figure 9.1 Parts of typical abutment. (a) = the bridge seat, which consists of pads on which rest the end bearings of the superstructure; (b) = the back or parapet wall, which affords lateral support for the upper part of the embankment; (c) = the main body of breast wall; (d) = the footings; (e) = the wing walls.

9.2.2 Straight-Wing Abutment

This is a special form of wing-wall abutment. In general, these are retaining walls modified so as to support the superstructure (Fig. 9.3). They are used with embankments of moderate height and mainly for smaller bridges. Straight-wing abutments are often built at stream crossings where the wings prevent the fill from blocking the stream, and where they tend to restrict the scouring action of water eddying around the main support.

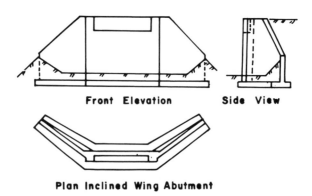

Figure 9.2 Wing-type abutment.

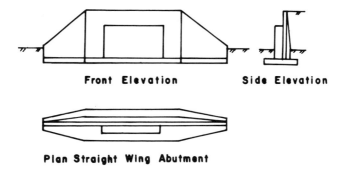

Figure 9.3 Straight-wing abutments.

This type is adaptable to use with skewed as well as square bridges. Such abutments are usually massive and must resist large overturning moments. The tops of the wings are sloped to conform to the slope of the embankment fill.

9.2.3 U-Type Abutments

The U abutment is a special form of wing-wall abutment in which the wings are parallel with the roadway (Fig. 9.4). Its use is especially suitable where rock slopes make possible stepping up the wing-wall footing to save masonry. This type should not be used in streams subjected to flooding, since there is a large amount of embankment outside of the wings that are not protected against scour. This type is not advantageous for crossings that are skewed sharply.

9.2.4 Box-Type Abutment

A box abutment is a partial box resting upon the ground. The roadway deck is an integral part of the abutment. A box abutment eliminates the fill that

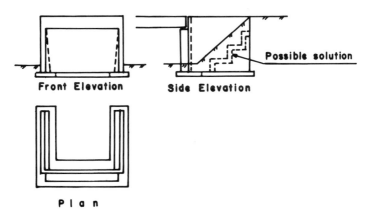

Figure 9.4 U-shaped abutment.

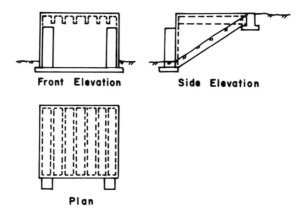

Figure 9.5 Box-type abutment.

required heavy retaining walls. It may, therefore, reduce the weight and thus be suitable for use on weaker soil while appearing to be massive (Fig. 9.5).

The two columns are practically piers on spread footings. The back wall behind the bearing is extended down as a curtain and may be utilized to help spread the bridge loads. The wings are also curtain walls that may or may not have footings. At the rear, a secondary wall and footing are supported upon undisturbed ground near the top of the slope, or they may be placed on piers, as shown by dotted lines, when the upper roadway is on an embankment.

It is usually best to build a box abutment as an integral structure. It may be skewed readily when the angle is not too sharp for proper appearance. It is a type worthy of study when the cut is deep or the fill is high.

9.2.5 Flanking-Span Abutment

It is logical in many cases to carry the tasks of reducing the lateral pressure on abutments and minimizing cost still further. Figure 9.6 shows a typical flanking span abutment. Here, the wing walls are omitted. When feasible, the front wall may also be reduced by using a rectangular or arched opening.

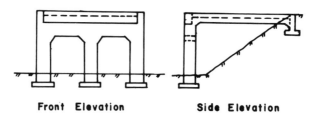

Figure 9.6 Flanking-span abutment.

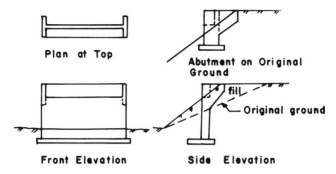

Figure 9.7 Abutments with small rear wings.

9.2.6 Floating Abutments

There are two types of wingless abutments:

1. *Abutments with Small Rear Wings* (Fig. 9.7). In this case, if the abutment is on the original ground of bearing value, rear wings are merely supported upon a shelf excavated in the side of the cut and then surrounded by backfill.

When the roadway is on an embankment, two or more piers may be used to support the bearing directly or to hold a long bridge seat. The wings may be very small, and they are perpendicular to the bridge seat or extended at the end parallel to it to keep the fill from the bearings, an important matter.

There is no attempt to terminate the bridge by means of any massive or special structure for aesthetic effect. The support is classified as an abutment only because it holds up the end of the superstructure.

2. *Abutments with Small Side Wings* (Fig. 9.8). This is another variation of wingless abutment.

9.3 MATERIAL

Modern bridge abutments are generally made of reinforced concrete.[5–7] Stonemasonry may be used occasionally, but it is ordinarily incorporated as a facing backed with concrete, especially when special architectural effects and durability of surface are desired.

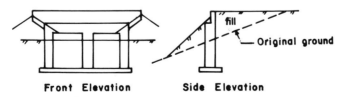

Figure 9.8 Abutments with small side wings.

164 SUBSTRUCTURE—ABUTMENTS

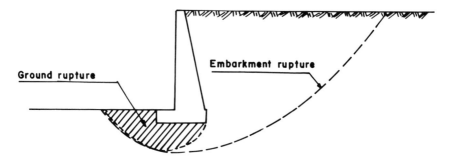

Figure 9.9 Stability of abutment.

9.4 DESIGN ANALYSIS

The forces considered for the design of piers apply to the design of abutments. The requirements for the classical design of abutments are:

1. To prevent uplift on the back side of the footing.

2. The resultant load of the lateral earth pressure and weight of the masonry and bridge must fall inside the middle third of the base. Live load should be considered.

3. For embankment stability it must be safe against the type of ground rupture shown in Figure 9.9. The direct cause of an embankment failure is insufficient depth of the abutment below the ground surface combined with the low shear strength of soil.

4. Proper drainage of the fill immediately behind the abutment is of primary importance. Water collecting behind the abutment not only increases the lateral earth pressure but alternate freezing and thawing materially reduces the durability of the concrete. Gravel fill, drains, and weep holes through the wall near the ground line are the elements of a good drainage system.

5. Because of the considerable difference in loadings, it is advisable to consider the wing walls as structures separated from the abutment and to provide expansion joints completely through walls and footings. However, many successful designs have been accomplished where abutments and wing walls are monolithic.

REFERENCES

1. Abbott, R. W., *American Civil Engineering Practice*, vol. III, Wiley, New York, 1957, pp. 36–32 to 26–41.
2. Kolokolov, N. M., Kopac, L. N., and Fainshtein, I. S., *Artificial Structures*, 3rd ed., Transport, Moscow, 1988, pp. 212–235 (in Russian).

3. Jacoby, H. S. and Davis, R. P., *Foundations of Bridges and Buildings*, 3rd ed., McGraw-Hill, New York, 1941, pp. 488–502.
4. Taylor, F. W., Thompson, S. E., and Smulski, E., *Concrete Plain and Reinforced*, Wiley, New York, 1950, pp. 406—422.
5. Gibshman, E. E., Kalmykov, N. J., Polivanor, N. I., and Kirillov, V. S., *Bridges and Structures on the Highways*, Autotransizdat, Moscow, 1961, pp. 441–448 (in Russian).
6. Platonov, E. V., *Supports of the Bridges*, 2nd ed., Transzheledorizdat, Moscow, 1946, pp. 490–517 (in Russian).
7. Mishchenko, A., Romantsov, Yu. V., and Yanovskii, G. M., *New Constructions of Bridge Abutments*, Transport, Moscow, 1987 (in Russian).

CHAPTER 10

AESTHETICS IN BRIDGE DESIGN

10.1 INTRODUCTION

The condition and character of bridges have been measures of civilization in all ages. As civilization developed, human needs increased and the desire for greater comfort created a need for transportation and communication. The bridge of fallen logs or swinging vines gave way to better and more commodious ones, over which loaded animals and carts could pass with safety. With the further advance of civilization and the extension of commerce, heavier and better bridges were required, until the coming of railroad transportation in the nineteenth century, when stronger ones were erected to carry trains.

The earliest bridges were for utility only, and little or no thought was given to their adornment. Primitive races were content with crude bridges that served only their barest needs. With the general development of culture there comes a stronger call for structures that please the eye and pay more attention to the aesthetic qualities.

The foundation of aesthetics is of the subjective order—the quality of the impression made on the mind of the observer by the thing observed. Therefore aesthetics are of a relative order and may gradually change with the developing mind. Artists and architects have formulated various principles during the past centuries defining their conceptions of the artistic. To these the engineer may often look for his first provisional standard for comparison.

Bridges are frequently the most conspicuous objects in the landscape. A bridge can often be seen for a great distance. No structure more clearly shows its objective and use, and the opportunity is therefore offered for truthful construction, a prime prerequisite for good design.

10.2 REQUIREMENTS FOR BRIDGE AESTHETICS

Bridges may be considered beautiful when they fulfill the requirements discussed below.[1-4]

10.2.1 Conformity with Environment

A bridge must conform with its surroundings and environment. The setting or surroundings greatly affect its appearance. Those that are exposed to the river view are seen and appreciated more than others that are partly hidden by adjoining objects.

Generally, a structure must be in harmony with its environment and not appear as an intrusion. To secure harmony between the structure and its environment means the merging of its general outlines with those of the landscape. It should be remembered that the bridge will likely be seen from various angles, and that each viewpoint will cause its own individual impression.

The rule, generally, is to make the bridge more striking than its surroundings, so the eye will naturally be attracted to it. A good method is to make separate photographs of the site and the design to the same scale, and after placing the proposed bridge in the landscape view, to rephotograph the combination. Features of the design that fail to conform to the surroundings will then appear, and changes can be made until it is satisfactory.

10.2.2 Economic Use of Material

Economic use of material is another standard of excellence. Beauty exists in every structure which is designed according to the principles of economy, with the greatest simplicity, the fewest members, and the most pleasing outline consistent with construction.

10.2.3 Exhibition of Purpose and Construction

The purpose of the bridge should be plainly evident, and generally the construction should be revealed. Expressiveness, to many people, is the main source of beauty. Strength and boldness should predominate.

10.2.4 Pleasing Outline and Proportions

A bridge is beautiful if its primary form or outline and its relative proportions are well and properly chosen. The proportions must satisfy the eye and the aesthetic feeling, and have optical harmony.[5-8] Good general lines are necessary as a basis; a consistent scale or proportion of parts should follow. A form that admits of no explanation cannot be beautiful. It must have and show some purpose in its general relation. Each part of a bridge structure should be treated

168 AESTHETICS IN BRIDGE DESIGN

in such a way that its function therein is apparent and emphasized according to its importance. The different kinds of materials used in structures call for different treatment and varying aesthetic standards.

The underlying thoughts connecting these precepts is that the structure must be fitted for the work it is to do, that it should express the truth, and that falsities are erroneous and outside the realm of rational aesthetics.

The achievement of good general lines is best attained by a study of the profile of the structure. The feature of a bridge is so pleasing to the eyes of all observers, cultivated and ignorant alike, as perfect symmetry in the layout of spans; consequently it should be attained whenever practical, even if some extra expense is involved thereby. Unfortunately, the connections are not always favorable to the perfect symmetry of design. In such cases it becomes necessary to do the best one can with the unfavorable conditions, and to make the structure aesthetically pleasing, if not symmetrical.

10.2.5 Appropriate but Limited Use of Ornament

Mere ornamentation generally affronts the sense of harmony and fitness. In many bridges what would otherwise be a pleasing outline is spoiled by the introduction of massive ornamental portals at the ends. It is not advisable to correct the hard, rigid outline of a span by the use of additional parts that falsely proclaim a different function for the members or confuse their action in the structure. Ornament is not architecture, and a bridge of beautiful outline may easily be spoiled with an excessive amount of details. Superfluous decoration has a minifying effect.

10.2.6 Expressiveness

Expressiveness in a structure is, for many people, the greatest element of beauty, and the visible parts and line show their purpose.[8-11] A bridge must be a truthful creation and its appearance should show its purpose. As strength is a chief requisite, the element should be emphasized.

10.2.7 Symmetry and Simplicity

One of the most important factors of good design is symmetry. If conditions permit, the general outline on each side of the center should be the same, or nearly so. The absence of symmetry should be permitted only when the ground contour or other conditions are such as to make a symmetrical arrangement impossible.

Simplicity is important though not so essential as symmetry. Too many members are confusing, and fewer large pieces are preferable. Simplicity means primarily a truth-telling structure, having no subterfuges about the line of stress, no covering of concrete or steel with a stone facing.

Symmetry may or may not be absolutely essential to a pleasing and satisfying design, although it is usually necessary if a truly artistic structure is to result. When the structure is of great length, unsymmetrical features are not so noticeable as in a short one, which will be seen at a glance. The most pleasing design is reached when the center of a bridge is at the center of the middle or main span.

10.2.8 Harmony and Contrast

Harmony is also essential to a pleasing design, because without it the structure would most surely be displeasing to the beholder. There must be harmony between the substructure and the superstructure; between the various component parts of the design, and complete harmony with the surroundings. Harmony is best exhibited where no parts of a structure or of its decorations seem to be extraneous and where the structure harmonizes with its surroundings.

To be a proper design a piece of engineering architecture must have symmetry or balance and harmony among its component parts and with its surroundings. Spans arranged in groups produce a better effect than a succession of similar ones, and groups should preferably contain three spans or more.

10.2.9 Material and Colors

The laws of harmony and contrast apply to the selection of material and colors. Heavy projections and deep shadows produce an effect of strength that is not easily secured without them. Color combinations produce harmony or discord on the senses. Soft colors are preferable to bright ones, and if two or more are used, they should, if possible, emphasize the construction lines.

The material from which an engineering work is to be constructed is a governing factor in its artistic treatment and what might be very appropriate for one kind of material would be very inappropriate for another.

10.2.10 Proportion

Proportion is closely related to each of the principles of design discussed above, and usually when the economic proportions have been determined the resulting designs are pleasing and artistic.

However, in most cases, modifications may be made so that the requirements of both economy and beauty can be satisfied. Proportion is most nearly reached when the entire structure is most pleasing to the trained eye, and the truth most closely adhered to in every part of the structure, and in every detail. The final design of the structure should only result after a most careful study of an outline elevation, with enough of the details sketched in, to make sure of the bearing of each item on the ensemble.

The designing of concrete and steel bridges must be governed by the foregoing principles, yet what is appropriate and harmonious for one type, may not be wholly satisfactory for another type of structure. Sincerity is more or less a matter of course in an engineering design, as the purpose of such a structure is nearly always evident and is covered by certain phrases of all the engineering postulates. Propriety is also to some extent covered by our basic principles, but mainly under the headings of symmetry and harmony. Style, as it is understood in architecture, is something that can seldom be considered in an engineering design, except insofar as an arch bridge might be considered Romanesque, or in other cases as somewhat Gothic. Scale is largely a feature of our principle of proportion, and to some extent it is related to symmetry and harmony.

10.3 CAUSES OF LACK OF AESTHETICS

The lack of beauty in bridges may result from:

1. *Indifference of designers and their lack of artistic training.* The designer's only objective has been to design a bridge of sufficient capacity and strength, and least cost.
2. *Competition and commercialism, resulting in the use of a contractor's plans.* Commercialism and competition are responsible to a great extent for a lack of art in bridges for, as a general rule, the cheapest bridge, and consequently the simplest one, was accepted.
3. *Lack of cooperation from architects.* Cooperation of architects was considered unnecessary. But where engineers and architects work together on bridges, the results should be fortunate.
4. *Absence of art standards for bridges.* Another reason for the lack of art is that no standards for bridges are available.
5. *Haste in construction.* Haste in construction is perhaps responsible for more ugly bridges than any other condition.
6. *Legal and financial obstacles.* A common explanation of unsightly bridges is the excuse of insufficient funds or appropriations.
7. *Inadequate material.* Suitable material is not always at hand, but this need not prevent the adoption of artistic forms for bridges.
8. *Unsuitable or unsymmetrical location.* The site also affects bridge appearance; if the surroundings are beautiful the bridge is more attractive. If the profile of the ground color is unsymmetrical, it is more difficult to make a symmetrical and satisfactory arrangement of spans.

10.4 AESTHETICS OF ORDINARY STEEL BRIDGES

Large steel bridges should always be proportioned according to the rules of economy and service, depending for their artistic effect on their general form.[12-14] A limited amount of ornament may be used on the balustrade, lamps, trolley pools, or brackets. Skew bridges should, if possible, be avoided.

10.4.1 Beam Bridges

Small spans are worthy of careful considerations and treatment, for they greatly outnumber the larger ones. Beam bridges are much used for street overpasses under railroad tracks where the latter are elevated on backs to avoid level crossing.

10.4.2 Truss Bridges

General principles of artistic design, such as symmetry and simplicity, should be applied whenever possible.

10.4.3 Movable Bridges

Some of the limitations are as follows:

- Number of decks and their relative position
- Elevation of deck above water
- Under clearance required
- Angle of crossing, whether square or skew
- Grade

Movable bridges show a greater lack of aesthetic treatment than almost any other form. Like other kinds, they must depend chiefly on their outline for their appearance, and their form should, so far as possible, show their purpose and action.

Each individual case requires different treatment, and a form which would be most suitable for one location might be quite unsuited to another. The surroundings of a structure greatly affect its appearance. Deck bridges are nearly always preferable to through ones, and should be used wherever enough height is obtainable beneath the floor for framing.

10.4.4 Cantilever Bridges

The cantilever or bracket bridge has merit peculiarly its own, but it is economical only when erection work would be very difficult or impossible. Structural requirements must always prevail, but it is no more difficult to make a cantilever attractive than a suspension or arch.

1. *Number of Spans.* There are three-span cantilevers and cantilevers with many spans.

2. *Chord Outline.* The cantilever, like other large steel bridges, should have a graceful outline if beauty is desired and curved chords are preferable artistically to straight ones. Curves may be used for either one or both chords, as conditions will allow. The center span bottom chord may be made a segment

of a circle and the bottom chord of the two adjoining anchor spans made to correspond with the middle one. Chord outlines resembling those of arches and suspensions are best suited to cantilevers that have no suspended span, for if such be introduced, the continuous curves produce less truss depth at the span center than at the ends—the reverse of the requirements.

10.4.5 Arches

Arches should exhibit character of their own. They contain three essential parts: (1) the floor, (2) the floor supports or spandrel framing, and (3) the arch ring.

1. *The Deck.* The deck should be arranged symmetrically with space for cars, vehicles, and pedestrians. A decided roadway camber is not only useful for drainage, but adds grace to the whole.

2. *Spandrel Framing.* Floor supports or spandrel framing of arches are similar to viaduct or trestle bends, and are similarly proportioned, the economic distance between columns depending on the height from arch to floor. A few larger bends are artistically preferable to a greater number of smaller ones. Economy is secured when flat arches with a small rise have a greater number of spandrel columns, but arches with a great rise should have those supports further apart. It should be noted that the relative elevation of roadway and springs give to arches their chief character.

3. *Arch Types.* The three common arch types are: (1) deck type; (2) through type; (3) tied arch. The form of arch also depends on the bearings, which may have either three, two, or no hinges. The three-hinged arch with joints at the ends and center must be stiff between those points, but may taper to a small depth at the bearings. For circular segments of arch forms, parabolas or hyperbolas have all been used, and any of them are suitable for arches of small rise, though for a condition approaching uniform loading, the parabola is nearest the line of pressure. Arches should have sufficient rise to display their strength.

10.4.6 Suspension Bridges

Suspension bridges achieve beauty with a minimum of effort, and can hardly fail to be attractive unless through deliberate purpose or negligence, for the cables naturally assume a perfect curve.[15, 16] The aesthetic appearance of bridges is greatly influenced by the number of spans. Suspension bridges generally have two towers, but they have been erected with only a single one, or with many towers.

No aesthetic treatment can be given to the cables themselves, for they are purely structural members. The curve taken by the unloaded cables alone is a catenary which has a most satisfying and pleasing appearance. The load of the roadway and trusses modifies this into a flat parabolic curve that differs very little in contour and appearance from that of a circular curve.

There are generally three types of suspension bridges, one type having a loaded center span with the backstays running direct to the anchorage; the second with loaded backstays, which makes the structure one of three spans; the third, with multiple spans, is self-anchored.

The backstays in the first type may be carried much flatter than the angle formed by a tangent to the parabolic curve and thus approximate the pleasing appearance of loaded backstays, which naturally produces the most artistic bridge.

The sag of cables has a very important bearing on the artistic appearance of a suspension bridge, as well as upon its economy and stiffness. When the sag is one tenth of the span, the bridge is much stiffer sidewise than for a greater depth, and the appearance is the best. Except in rare cases, the sag should not be made as small as one twelfth of the span nor deeper than one ninth.

The stiffening trusses, which usually are suspended from the cables, are perhaps most pleasing when they are approximately straight, following the cambered curve of the roadway. The towers of a suspension bridge are its most prominent features and afford the greatest opportunity for artistic treatment. The design of above ground anchorages have been much more successful, and very often they are beautifully detailed masonry structures, with finely designed and attractive parapets.

10.4.7 Cable-Stayed Bridges

The introduction of the cable-stayed system is a true pioneering development in bridge architecture.[17-19] Existing cable-stayed bridges are masterpieces of steel and concrete construction. They are pleasing in outline, clean in their anatomical conception, and totally free of meaningless ornamentation. This is because the design of cable-stayed bridges was governed not only by financial, practical, and technical requirements, but also, to a great extent, by aesthetic and architectural considerations. In the design of modern cable-stayed bridges, one objective is to produce an aesthetically appealing bridge that blends with its surroundings. These bridges are truly representative of modern times. They are the product of engineering science, which is consistently advancing in accordance with its own laws and has been given form and substance by the twentieth-century engineer.

In span after span designers have demonstrated that beauty in cable-stayed bridges can be obtained without sacrificing either utility or economy. The new, slender and elegant bridge forms were created with steel, concrete, and mathematics in a combination of strength and beauty.

From an aesthetic point of view, cable-stayed bridges have a pleasing shape, as they clearly reveal the function of the cables and towers, and because the cables, owing to their small perimeter, are very unobtrusive to the overall appearance of the bridge.

It should be noted that, at the present time, the structural system of cable-stayed bridges in their numerous variants is coming into prominence. The thinking and the daring exemplified by these bridges constitute the most important contributions to the development of modern bridge engineering.

The most striking feature about cable-stayed bridges built during the last decade is not their technical features, though without them their new look would not be possible, but their elegant form and the way in which they have been blended with the landscape. Some of these notable bridges have been constructed by engineers in close collaboration with architects and town planners. Each in his own sphere has striven to give his best in a joint venture, so that as a team they have been able to achieve a work that excels that which each would have contributed individually.

Some features of cable-stayed bridges are:

1. Order—a harmony of scale, material, and detail, considering elements and the whole structure.
2. Arrangement—or expression of function, which is the basis of good design.
3. Eurythmy—or the element of beauty resulting from the fitting adjustment of members.
4. Symmetry—or proper balance of parts, considering the ratio of height of opening to span and of width of structure to span.
5. Propriety—or honesty in design and lack of deception. The best bridge design can always have a certain simplicity.
6. Economy—considering cost, efficiency, and lack of superfluity.

A comparison of cable-stayed bridges indicates that a layout of the radial system does not always provide the most attractive solution, especially where a double-space system is employed. The harp arrangement of parallel cables is preferable here, as the unsightly intersections in the lines of cables by the radial cable arrangement are avoided.

Towers are the most conspicuous part of the bridge, being visible from many points on land and river. It is therefore important to give them appropriate aesthetic treatment. The guiding motives usually are structural simplicity and harmonious proportions. The height is determined by the clear height of the deck above water as the center, the depth of the deck structure, and the selected arrangement and inclination of the cables. The surfaces of the towers may be given a pleasing appearance by an appropriate choice of form and the constructing of their sections economically, using only metal required for structural reasons. Architecturally, free-standing towers have a pleasant appearance, especially when only two towers are used along the longitudinal center line of the bridge. This system is particularly attractive because in elevation there is no intersection of the lines of the cables, and the road user has an obstructed view from the bridge on one side.

The aesthetic effect produced by cable-stayed bridges may be better understood by further analysis of a few outstanding bridges that utilize this new bridge form.

10.4.8 Reinforced Concrete Bridges

Reinforced concrete spans are generally shorter in comparison with steel. The length of span is usually determined by local conditions. Bridges of many spans appear best when the center one is longer than the others, and adjoining ones decrease in length toward the ends. Reinforced concrete bridges should have their decks symmetrically and carefully arranged, with enough space for traffic. Spandrels are of two kinds, solid and open.

Shape and Proportion. The appearance of reinforced concrete bridges depends chiefly on the arch curve, and a form should be selected that is the most pleasing consistent with construction. The common forms are (a) the semicircle; (b) the ellipse; and (c) the circular segment.

The first is preferable for long series of arches or high viaducts. The semicircle and ellipse are always satisfying, the ellipse being merely an oblique view of the circle. For comparatively flat arches, a curve of the same rise halfway between a segment and an ellipse corresponds closely with the line of pressure, but departure from exact curves produces optical discord. The segment is the correct constructive form for exhibiting greatest strength, but the semicircle and ellipse are acceptable for their fine appearance. Ellipses seem to be weak when they have too small a rise, the flat central part contrasting with the greater curvature of the springs. Segmental and elliptical arches appear to best advantage on low bridges, for the form originated from insufficient space for a greater rise.

REFERENCES

1. Tyrell, H. G., *Artistic Bridge Design*, Myron C. Clark, Chicago, 1912, p. 20.
2. Fowler, C. E., *The Ideals of Engineering Architecture*, Gillette, Technical Publications, Chicago, IL, 1929.
3. Vitruvius, *The Ten Books on Architecture*, translated by M. H. Morgan, Dover Publications, New York, 1960, pp. 13–16.
4. Torroja, E., *Philosophy of Structures*, University of California Press, Berkeley and Los Angeles, 1962, pp. 268–269.
5. Hartmann, F., *Aesthetic in Bridge Engineering*, Franz Deuticke, Leipzig and Wien, 1928 (in German).

6. Pacholik, L., *Aesthetics of Bridge Structures*, Grafic Publishing House, Prague, 1946 (in Czechoslovakian).
7. Démaret, J., *Aesthetic and Construction of the Structures*, Dunode, Paris, 1948 (in French).
8. Ministry of Transport, *The Appearance of Bridges*, Her Majesty's Stationery Office, London, 1969.
9. The Institution of Civil Engineers, The aesthetic aspect of civil engineering design, A record of six lectures delivered at the institution, 1955.
10. Mock, E. B., *The Architecture of Bridges*, Museum of Modern Art, New York, 1948.
11. Mays, R. R., Beautiful bridges, *Civ. Eng.*, *ASCE*, 72–74 (August 1989).
12. Backow, A. F. and Kruckemeyer, K. E. (ed.), *Bridge Design—Aesthetics and Developing Technologies*, Massachusetts Department of Public Works, 1986.
13. Shchusev, P. V., *Bridges and Their Architecture*, State Edition on Construction and Architecture, Moscow, 1952 (in Russian).
14. Waddell, J. A. L., *Bridge Engineering*, vol. II, Wiley, New York, 1916, pp. 1150–1181.
15. Billington, D. P., *The Tower and the Bridge*, Basic Books, New York, 1983.
16. Steinman, D. B., *A Practical Treatise on Suspension Bridges*, 2nd ed., Wiley, New York, 1953.
17. Podolny, W. Jr. and Scalzi, J. B., *Construction and Design of Cable-Stayed Bridges*, 2nd ed., Wiley, New York, 1986.
18. Troitsky, M. S., *Cable-Stayed Bridges, An Approach to Modern Bridge Design*, 2nd ed., Van Nostrand Reinhold, New York, 1988, pp. 36–41.
19. Gimsing, N. J., *Cable Supported Bridges, Concept and Design*, Wiley, New York, 1983.

CHAPTER 11

SPECIFICATIONS AND CODES

11.1 GENERAL DATA

The following are common specifications for steel and concrete bridges in the United States and Canada:

1. American Association of State Highway and Transportation Officials, *Standard Specifications for Highway Bridges*, 15th ed., 1992, and *Interim Specifications*, 1992.
2. *Design of Highway Bridges*, National Standard of Canada, CAN/CSA-S6-88.
3. *Ontario Highway Bridge Design Code*, 1979, vols. 1 and 2, Ontario Ministry of Transportation and Communication.
4. *AASHTO Manual on Foundation Investigation*, American Association of State Highway and Transportation Officials.
5. American Association of State Highway and Transportation Officials, *Guide Specifications and Commentary for Vessel Collision Design of Highway Bridges*, volume I: Final Report, February 1991.
6. American Association of State Highway and Transportation Officials, *Standard Specifications for Movable Highway Bridges*, 1988.
    a. Proposed Design Specifications for Steel Box Girder Bridges, Report No. FHWA-TS-80-205, January 1980, Final Report, Federal Highway Administration.
    b. Design Examples for Steel Box Girders, Report No. FHWA-TS-86-209, July 1986, Final Report, Federal Highway Administration.

7. Recommendations of Stay Cable Design, Testing and Installation, Post-Tensioning Institute, February 1990.

There are also the specifications that are related to steel bridges only, as follows:

1. Specifications for Steel Railway Bridges, AREA, Chicago, 1969.
2. ANSI/AASHTO/AWS Bridge Welding Code D1.5-88, 1988.
3. General Specification for Welding of Steel Structures, CSA Standard W59.1-1970, Toronto, 1970.
4. Research Council on Riveted and Bolted Structural Joints of the Engineering Foundation. "Specification for Structural Joints Using ASTM A325 Bolts," *Proc. Am. Soc. Civ. Eng.*, **88** (ST-5), 11 (October 1962).
5. O. G. Julian "Synopsis of First Progress Report of Committee on Factors of Safety," *Proc. Am. Soc. Civ. Eng.*, **83** (ST-4) 1316 (July, 1957).

The following specifications are related to the reinforced and prestressed concrete bridges only:

1. Concrete Members with Variable Moment of Inertia, ST 103, Portland Cement Association.
2. Beam Factors and Moment Coefficients for Members with Prismatic Haunches, ST 81, Portland Cement Association.
3. Beam Factors and Moment Coefficients for Members with Intermediate Expansion Hinges, ST 75, Portland Cement Association.
4. Frame Analysis Applied to Flat Slab Bridges, ST 64-2, Portland Cement Association, 1944.
5. *Reinforced Concrete Design Handbook Working Stress Method SP-3*, American Concrete Institute, 3rd ed., 1980.
6. Notes on Load Factor Design for Reinforced Concrete Bridge Structures with Design Applications, Portland Cement Association.
7. ACI Building Code Requirements for Reinforced Concrete (ACI 318 89) and Commentary, American Concrete Institute.
8. Strength and Serviceability Criteria, Reinforced Concrete Bridge Members, Ultimate Design, FHWA, October 1969.
9. Continuous Hollow Girder Concrete Bridges, Portland Cement Association.
10. Continuous Concrete Bridges, Portland Cement Association.
11. *Post-Tensioned Box Girder Bridge Manual*, Post Tensioning Institute, 1978.
12. *Post-Tensioned Box Girder Bridges—Design and Construction*, Western Concrete Reinforcing Steel Institute, 1969.

13. *Post-Tensioning Manual*, Post Tensioning Institute, 1976.
14. *Precast Segmental Box—Girder Bridge Manual*, Prestressed Concrete Institute, 1978.
15. American Association of State Highway and Transportation Officials, *Guide Specifications for Design and Construction of Segmental Concrete Bridges*, 1989.
16. J. G. MacGregor, U. H. Oethafen, and S. E. Hage, *A Re-examination of the El Value for Slender Columns*, ACI Publication SP-50-1, Reinforced Concrete Columns, 1975.

In addition to specifications, designs are frequently controlled by local, regional, or national codes. These codes sometimes incorporate specifications either by direct reference, by incorporation of all or part of the specifications, or by rewriting and revision to suit their particular needs.

It should be noted that the most recent specification changes show a significant departure from past practice. It is a trend that will continue with new developments in materials, in fabrication, and in design concepts and approaches. The effect of these revisions is important. They provide more attention to matters of economy, through selection of materials with the most appropriate strength, and other characteristics such as choice among bolts and welding. A mixture of steels may be contemplated in design. Although approximations for rapid design use remain, specifications are permitting more and more design refinements. In some cases, these refinements necessarily involve more complex procedures, but developments in the use of computers have made such rapid strides that the matter of complexity is not the problem it once was. In the long run, the improved procedure leads to more logical designs.

11.2 LOADS ON BRIDGES

11.2.1 Dead Loads

The dead load acting on a bridge is a sum of the weights of all its components. If the bridge cross-section does not change radically along the length of the bridge, the dead load may be considered as uniformly distributed and the dead weight of the entire structure is carried by the main members.

The designer must include all other loads carried by the bridge, such as those applied by sidewalks and utilities. An allowance of 10–15% is sometimes applied for such details as bolts, rivets, appurtenances, gussets, and paintwork.

Some typical values useful in dead weight calculations are:

Steel	490 lb/ft^3
Concrete (plain or reinforced)	150 lb/ft^3

Ballast	120 lb/ft^3
Loose sand	100 lb/ft^3
Macadam	140 lb/ft^3
Railway rails, guard rails, fastening	200 lb/ft^3
Aluminum	170 lb/ft^3
Timber	35–60 lb/ft^3
Asphalt paving	140 lb/ft^3
Cast iron	450 lb/ft^3

11.2.2 Live Loads

The present design vehicles contained in the AASHTO specifications were adopted in 1944. The loading consists of five weight classes, namely: H10, H15, H20, HS15, and HS20. The design vehicles for each of the five classes are shown in Figures 11.1 and 11.2. Any actual vehicle that would be permitted to cross a bridge should not produce stresses greater than those caused by these hypothetical vehicles.

The lighter loads, H10 and H15, are used for the design of lightly traveled roads, while the H20 and HS20 are used for national highways. The HS20 is used for the design of bridges on the interstate highway system. An additional alternate loading was introduced for this system. This loading consists of two axles spaced at 4 ft and weighing 24 k each.

The HS truck loading shows a variable spacing of two rear axles from 14 to 30 ft. The correct spacing is the length that produces the maximum effect. For stresses in simple span bridges, the spacing is the minimum value of 14 ft. However, for continuous spans, a spacing greater than 14 ft may produce the maximum stresses. The influence diagram indicates the proper spacing of the axles which result in maximum stresses. In addition to the truck loading, the specifications contain equivalent loadings (Fig. 11.3) to be used in place of the truck loadings when they produce a greater stress than the truck.

Prior to the 1944 specifications, the design live load consisted of the basic H trucks preceded and followed by a train of trucks weighing three-quarters as much as the basic truck. In 1944 the HS truck was developed and the equivalent lane loading took the place of the train of trucks. Presently only one truck is to be used per lane per span. For longer spans the equivalent loading produces greater stresses than the single truck. The span length at which the loading changes for shear calculations is different than that for moments calculations.

The concentrated load used in the equivalent lane loadings is different for moment than for shear calculations. Only one concentrated load is used in a simple span for a positive moment in continuous spans. Two concentrated loads are used for the determination of the negative moments. The equivalent lane load is placed so as to produce the maximum stresses. The uniformly distributed load can be divided into segments when applied to continuous spans. Both the concentrated load and the uniform load are distributed over a 10-ft lane width on a line normal to the center line of the lane.

11.2 LOADS ON BRIDGES

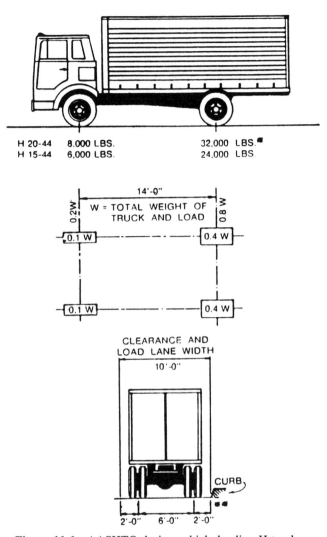

Figure 11.1 AASHTO design vehicle loading H trucks.

11.2.3 Reduction in Load Intensity

Where maximum stresses are produced in any member by loading a number of traffic lanes simultaneously, the following percentages of the live loads are used in view of the improbability of coincident maximum loading:

Lanes	Percent
One or two	100
Three	90
Four or more	75

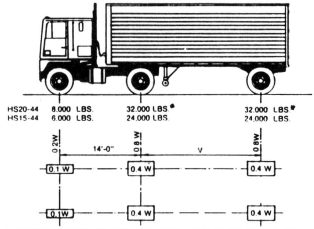

Figure 11.2 AASHTO design vehicle loading HS trucks.

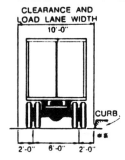

Figure 11.3 AASHTO design lane loading.

11.2.4 Sidewalk Loading

Sidewalk floors, stringers, and their intermediate supports are designed for a live load of 85 lb/ft² of sidewalk area. Girders, trusses, arches, and other members are designed for the following sidewalk live loads:

$$\text{Spans 0-25 ft in length} \qquad 85 \text{ lb/ft}^2$$
$$\text{Spans 26-100 ft in length} \qquad 60 \text{ lb/ft}^2$$

Spans over 100 ft in length are designed according to the formula

$$P = \left(30 + \frac{3000}{L}\right)\left(\frac{55 - W}{50}\right) \qquad (11.1)$$

in which

P = live load per square foot, max. 60 lb/ft²
L = loaded length of sidewalk (ft)
W = width of sidewalk (ft)

11.2.5 Impact

The amount of impact allowance or increment is expressed as a function of the live load stress, and shall be determined by the formula

$$I = \frac{50}{L + 25} \qquad (11.2)$$

in which

I = impact factor (maximum 30%)
L = length in feet of the portion of the span that is loaded to produce the maximum stress in the member

For uniformity of application, in this formula the loaded length L is as follows:

1. For roadway floors: the design span length.
2. For transverse members, such as floor beams: the span length of member center to center of supports.
3. For computing truck loads moments: the span length, or for cantilever arms, the length from the moment center to the farthermost axle.
4. For shear due to truck load: the length of the loaded portion of span from the point under consideration to the far reaction: except, for cantilever arms, use a 30% impact factor.

5. For continuous spans: the length of span under consideration for the positive moment and the average of two adjacent loaded spans for the negative moment.

11.2.6 Longitudinal Forces

Provision shall be made for the effect of a longitudinal force of 5% of the live load in the all lanes carrying traffic headed in the same direction. All lanes shall be loaded for bridges likely to become one-directional in the future. The load used without impact is the lane load plus the concentrated load for moment, with reduction for multiple-loaded lanes. The center of gravity of the longitudinal force is assumed to be located 6 ft above the floor slab and transmitted to the substructure through the superstructure. Provision is also made for the longitudinal forces due to friction at expansion bearings or shear resistance elastometric bearings.

11.2.7 Wind Loads

The wind load consists of moving uniformly distributed loads applied to the exposed area of the structure. The exposed area is the sum of the area of all members, including floor systems and railing, as seen in elevation at 90 degrees to the longitudinal axis of the structure. The forces and loads given herein are for a base wind velocity of 100 mph. For Group II and Group V loadings, but not for Group III and Group VI loadings, they may be reduced or increased in the ratio of the square of the design wind velocity to the square of the base wind velocity provided that maximum probable wind velocity can be ascertained with reasonable accuracy, or provided that there are permanent features of the terrain that make such changes safe and advisable. A change in the specification design wind velocity should be shown on the plans.

Superstructure Design

1. *Group II and Group IV Loadings.* A wind load of the following intensity is applied horizontally at right angles to the longitudinal axis of the structure:

For trusses and arches	75 lb/ft^2
For girders and beams	50 lb/ft^2

The total force shall not be less than 300 lb/linear ft in the plane of the windward chord and 150 lb/linear ft in the plane of the leeward chord on truss spans, and not less than 300 lb/linear ft on girder spans.

2. *Group III and Group VI Loadings.* Group III and Group IV loadings shall compromise the loads used for Group II and Group IV loadings reduced by 70% and a load of 100 lb/linear ft applied at right angles to the longitudinal axis of the structure and 6 ft above the deck as a wind load on a moving live load. When a reinforced concrete floor slab or a steel grid deck is keyed to or

attached to its supporting members, it may be assumed that the deck resists, within its plane, the shear resulting from the wind load on the moving live load.

Substructure Design

Forces transmitted to the substructure by the superstructure and forces applied directly to the substructure by wind loads are as follows: the transverse and longitudinal forces transmitted by superstructure to the substructure for various angles of wind direction are set forth in Table 11.1.

The skew angle is measured from the perpendicular to the longitudinal axis and the assumed wind direction produces the maximum stress in the substructure. The transverse and longitudinal forces are applied simultaneously at the elevation of the center of gravity of the exposed area of the superstructure.

The loads tested above are used in Group II and Group IV loadings. For the usual girder and slab bridges having maximum span length of 125 ft, the following wind loading may be used in lieu of the more precise loading specified above:

W—Wind load of structure
   50 lb/ft^2 transverse
   12 lb/ft^2 longitudinal
   Both forces shall be applied simultaneously.

WL—Wind load on live load
   100 lb/linear ft, transverse
   40 lb/linear ft, longitudinal
   Both forces shall be applied simultaneously

1. *Forces Applied Directly to the Substructure.* The transverse and longitudinal forces to be applied directly to the substructure for a 100 mph wind are calculated from an assumed wind force of 40 lb/ft^2. For wind directions assumed skewed to the substructure this force shall be resolved into components perpendicular to the end and from elevations of the substructure. The component perpendicular to the end elevation shall act on the exposed substructure area as seen in end elevation and the component perpendicular to the front

TABLE 11.1 Skewed Wind Forces Applied to the Substructure

	Trusses		Girders	
Skew Angle of Wind (deg)	Lateral Load (PSF)	Longitudinal Load (PSF)	Lateral Load (PSF)	Longitudinal Load (PSF)
0	75	0	50	0
15	70	12	44	6
30	65	28	41	12
45	47	41	33	16
60	24	50	27	19

elevation shall act on the exposed areas and shall be applied simultaneously with the wind loads from the superstructures. The loads given above are for Group II and Group V loading and may be reduced by 70% for Group III and Group VI loadings.

2. *Overturning Forces.* The effect of forces tending to overturn structures are calculated under Groups II, III, and VI, assuming that the wind direction is at right angles to the longitudinal axis of the structure. In addition, an upward force is applied at the windward quarter point of the transverse superstructure width. This force is 20 lb/ft^2 of deck and sidewalk plan area for Group II and Group IV combinations and 6 lb/ft^2 for Group III and Group IV combinations.

11.2.8 Thermal Forces

Provisions are made for stresses of movements resulting from variations of temperature. The rise and fall in temperature is fixed for the locality in which the structure is to be constructed and computed from an assumed temperature at the time of erection. Due consideration shall be given to the lag between air temperature and the interior temperature of massive concrete members or structures. The range of temperature is generally as follows:

Metal Structures:
  Moderate climate, from 0 to 120°F
  Cold climate, from −30 to 120°F.

Concrete Structures:

	Temperature Rise (°F)	Temperature Fall (°F)
Moderate climate	−30	−40
Cold climate	−35	−45

11.2.9 Uplift

Provision is made for adequate attachment of the superstructure to the substructure by ensuring that the calculated uplift to any support is resisted by tension members engaging a mass of masonry equal to the largest force obtained under one of the following conditions:

a. 100% of the calculated uplift caused by any loading or combination of loadings in which the live plus impact loading is increased by 100%.
b. 150% of the calculated uplift at working load level. Anchor bolts subject to tension or other elements of the structure stressed under the conditions described above are designed at 150% of the allowable basic stress.

11.3 DISTRIBUTION OF LOADS TO STRINGERS, LONGITUDINAL BEAMS, AND FLOOR BEAMS

11.3.1 Position of Loads for Shear

In calculating end shears and end reactions in transverse floor beams and longitudinal beams and stringers, no longitudinal distribution of the wheel load is assumed for the wheel or axle load adjacent to the end of which the stress is being determined. Lateral distribution of the wheel load is that produced by assuming that the flooring acts as a simple span between stringers or beams. For loads in other positions on the span, the distribution for shear is determined by the method prescribed for moment.

11.3.2 Bending Moments in Stringers and Longitudinal Beams

In calculating bending moments in longitudinal beams or stringers, no longitudinal distribution of the wheel loads is assumed. The lateral distribution is determined as follows.

The live-load bending moment for each interior stringer is determined by applying to the stringer the fraction of a wheel load (both front and rear) determined in Table 11.2.

The dead load supported by the outside roadway stringer or beam is the portion of the floor slab carried by the stringer or beam. Curbs, railing and wearing surface, if placed after the slab has cured, may be distributed equally to all roadway stringers or beams. The live-load bending moment for outside roadway stringers or beams is determined by applying to the stringer or beam the reaction of the wheel load obtained by assuming the flooring to act as a simple span between stringers or beams. In no case shall an exterior stringer have less carrying capacity than an interior stringer.

In the case of a span with a concrete floor supported by four or more steel stringers, the fraction of the wheel load shall not be less than

$$\frac{S}{5.5} \tag{11.3}$$

TABLE 11.2 Distribution of Wheel Loads in Longitudinal Beams (Abbreviated)

Kind of Floor	Bridge Designed for One Traffic Lane	Bridge Designed for Two or More Traffic Lanes
Concrete on steel I-beam stringers and prestressed concrete girders	$S/7.0$	$S/5.5$
On concrete T-beam	$S/6.5$	$S/6.0$
Concrete box girders	$S/8.0$	$S/7.0$

TABLE 11.3 Distribution of Wheel Loads in Transverse Beams (Abbreviated)

Kind of Floor	Fraction of Wheel Load to Each Floor Beam[a]
Concrete	$Sf/6$
Steel grid (less than 4 in. thick)	$Sf/4.5$
Steel grid (4 in. or more)	$Sf/6$

[a] If S exceeds denominator, the load on the beam is the reaction of the wheel loads assuming the flooring between beams acts as a simple beam.

where $S = 6$ ft or less and is the distance in feet between outside and adjacent interior stringers, and where S is more than 6 ft and less than 14 ft.

$$\frac{S}{4.0 + 0.25S} \quad (11.4)$$

When S is 14 ft or more the load on each stringer is the reaction of the wheel load, assuming the flooring between the stringers acts as a simple beam.

Bending Moments in Floor Beams (Transverse). In calculating bending moments in floor beams, no transverse distribution of the wheel loads is assumed. If longitudinal stringers are omitted and the floor is supported directly on floor beams, the beams shall be designed for loads determined in accordance with Table 11.3.

11.4 SUBSTRUCTURE

11.4.1 Forces from Stream Current, Floating Ice, and Drift

All piers and other portions of structures that are subject to the force of flowing water, floating ice, or drift are designed to resist the maximum stresses induced thereby.

1. *Force of Stream Current on Piers.* The effect of flowing water on piers is calculated by the formula

$$P = KV^2 \quad (11.5)$$

where

P = pressure (lb/ft^2)
V = velocity of water (fps)

K = a constant being $1\frac{3}{8}$ for square ends, $\frac{1}{2}$ for angle ends where the angle is 30 degrees or less, and $\frac{2}{3}$ for circular piers

2. *Force of Ice on Piers.* Ice forces on piers are selected with regard to site conditions and the mode of ice action to be expected. Consideration is given to the following modes:

   a. Dynamic ice pressure due to moving ice sheets and ice floes carried by stream flow, wind, or currents.

   b. Static ice pressure due to thermal movements of continuous stationary ice sheets on large bodies of water.

   c. Static pressure resulting from ice jams.

   d. Static uplift or vertical loads resulting from adhering ice in waters of fluctuating level.

3. *Dynamic Ice Force.* Horizontal forces resulting from the pressure of moving ice is calculated by the formula:

$$F = C_n p t w \qquad (11.6)$$

where

F = horizontal ice force on pier (lb)
C_n = coefficient for nose inclination from Table 11.4
p = effective ice strength (psi)
t = thickness of ice in contact with pier (in.)
w = width of pier or diameter of circular-shaft pier at the level of ice action (in.)

The effective ice strength p is normally in the range of 100 to 400 psi on the assumption that crushing or splitting of the ice takes place on contact with the pier. The following values of effective ice strength appropriate to various situations may be used as a guide:

   a. On the order of 100 psi where breakup occurs at melting temperatures and where the ice runs as small "cakes" and is substantially disintegrated in its structure.

   b. On the order of 200 psi where breakup occurs at melting temperature but the ice moves in large pieces and is internally sound.

TABLE 11.4 Coefficient for Nose Inclination

Inclination of Nose to Vertical	C_n
0–15	1.00
15–30	0.75
30–45	0.50

c. On the order of 300 psi where at breakup there is an initial movement of the ice sheet as a whole or where large sheets of sound ice may strike the piers.

d. On the order of 400 psi where breakup or major ice movement may occur with ice temperature significantly below the melting point.

e. The preceding values for effective ice strength are intended for use with piers of substantial mass and dimensions. The values are modified as necessary for variations in pier width or pile diameter, and design ice thickness by multiplying by the appropriate coefficient obtained from Table 11.4.

f. *Static ice pressure.* Ice pressure on piers frozen into ice sheets on large bodies of water receive special consideration when there is reason to believe that the ice sheets are subject to significant thermal moments relative to the piers.

4. *Buoyancy.* Buoyancy is considered where it affects the design of either the substructure, including piling, or the superstructure.

11.5 EARTH PRESSURE

1. Structures that retain fills are proportional to withstand pressure as given by Rankine's formula provided, however, that no structure is designed for less than an equivalent fluid weight (mass) of 30 lb/ft^3.

2. For rigid frames a maximum of one-half of the moment caused by earth pressure (lateral) may be used to reduce the positive moment in the beams, in the top slab, or in the top and bottom slab.

3. When highway traffic can come within a horizontal distance from the top of the structure equal to one-half its height, the pressure has a live-load surcharge pressure equal to not less than 2 ft of earth added to it.

4. Where an adequately designed reinforced concrete approach slab supported at one end by the bridge is provided, no live-load surcharge need be considered.

5. All design shall provide for the thorough drainage of the back-filling material by means of weep holes and crushed rock, pipe drains or gravel drains, or perforated drains.

11.6 SEISMIC DESIGN

In regions where earthquakes may be anticipated, structures are designed to resist earthquakes following AASHTO *Standard Specifications for Highway Bridges*, 15th ed., 1992, Division I-A Seismic Design, pp. 333 to 433.

11.6 SEISMIC DESIGN

11.6.1 Flow Charts and Example for Use of Standards

Note: Articles indicated on Figures 11.4 and 11.5 refer to the AASHTO *Standard Specifications for Highway Bridges*, 15th ed., 1992, Sections 3, 4, 6, 7, and 8.

11.6.2 Applicability of Standards

These standards are for the design and construction of new bridges to resist the effect of earthquake motions. The provisions apply to bridges of conventional steel and concrete girder and box girder construction with spans not exceeding 500 ft (152.4 m). Suspension bridges, cable-stayed bridges, arch type, and movable bridges are not covered by these standards but general considerations for designing such bridges are presented in the commentary.

No detailed seismic analysis is required for any single-span bridge or for any bridge in seismic Performance Category A. For both single-span bridges (Article 4.5), bridges classified as SPCA (Article 4.5), and bridges classified as SPCA (Article 4.6) the connections must be designed for specified forces and must also meet minimum support length requirements (Article 4.9).

11.6.3 Section 3. General Requirements

    3.1   Applicability of Standards
    3.2   Acceleration Coefficient
    3.3   Importance of Classification
    3.4   Seismic Performance Categories
    3.5   Site Effects
    3.6   Response Modification Factors

11.6.4 Section 4. Analysis and Design Requirements

    4.1   General
    4.2   Analysis Procedure
    4.3   Determination of Elastic Forces and Displacements
    4.4   Combination of Orthogonal Seismic Forces
    4.5   Design Requirements for Single-Span Bridges
    4.6   Design Forces for Seismic Performance Category A
    4.7   Design Forces for Seismic Performance Category B
    4.8   Design Forces for Seismic Performance Categories C and D
    4.9   Design Displacements

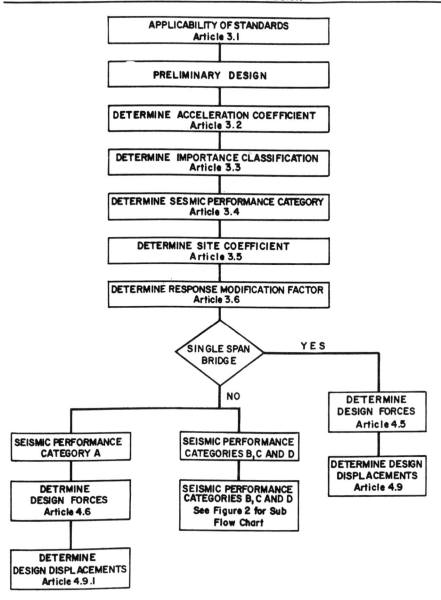

Figure 11.4 Design procedure flowchart.

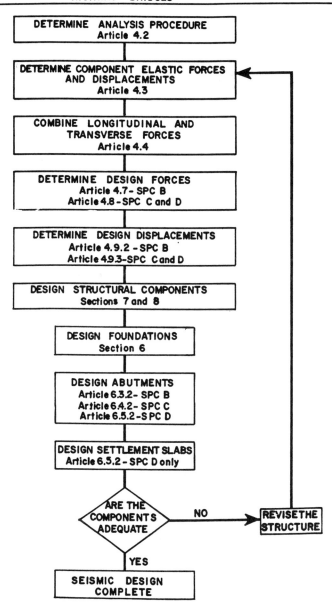

Figure 11.5 Sub-flowchart for seismic performance categories B, C, and D.

11.6.5 Section 5. Analysis Methods

    5.1   General
    5.2   Elastic Seismic Response Coefficient and Spectrum
    5.3   Single Mode Spectral Analysis Method—Procedure 1
    5.4   Multimode Spectral Analysis Method—Procedure 2

11.6.6 Section 6. Foundation and Abutment Design Requirements

    6.1   General
    6.2   Seismic Performance Category A
    6.3   Seismic Performance Category B
    6.4   Seismic Performance Category C
    6.5   Seismic Performance Category D

11.6.7 Section 7. Structural Steel

    7.1   General
    7.2   Seismic Performance Category A
    7.3   Seismic Performance Categories B, C, and D

11.6.8 Section 8. Reinforced Concrete

    8.1   General
    8.2   Seismic Performance Category A
    8.3   Seismic Performance Category B
    8.4   Seismic Performance Categories C and D

COMMENTARY

Section 1	Introduction
	Design Philosophy
	Seismic Ground Motion Accelerations
	Soil Effects on Ground Motion

COMMENTARY

Section 3	General Requirements
C.3.1	Applicability of Standards
C.3.2, C.3.5, and C.5.2	Acceleration Coefficient, Site Effects and Elastic Seismic Response Coefficient and Spectrum

C.3.3	Importance Classification
C.3.4	Seismic Performance Categories
C.3.5	Site Effects
C.3.6	Response Modifications Factors

COMMENTARY

Section 4	Analysis and Design Requirements
C.4.1	General
C.4.2	Analysis Procedure
C.4.3	Determination of Elastic Forces and Displacements
C.4.4	Combination of Orthogonal Seismic Forces
C.4.5	Design Requirements for Single Span Bridges
C.4.6	Design Forces for Seismic Performance Category A
C.4.7	Design Forces for Seismic Performance Category B
C.4.8	Design Forces for Seismic Performance Categories C and D

COMMENTARY

Section 5	Analysis Procedures
C.5.1	General
C.5.2	Elastic Seismic Response
C.5.3	Single Mode Spectral Analysis Method
C.5.4	Method Multimode Spectral Analysis

COMMENTARY

Section 6	Foundation and Abutment Design Requirements
C.6.3, C.6.4 and, C.6.5	Seismic Performance Categories B, C, and D
C.6.3.2, C.6.4.2, and C.6.5.2	Abutments

COMMENTARY

Section 7	Structural Steel
C.7.1	General
C.7.3	Seismic Performance Categories B, C, and D

COMMENTARY

Section 8	Reinforced Concrete
C.8.1	General
C.8.2	Seismic Performance Category A
C.8.3	Seismic Performance Category B
C.8.4	Seismic Performance Categories C and D

Supplement A Work example using the standards. This example is intended to illustrate the application of the standards. The bridge selected for the example is a three-span, continuous box-girder structure with dimensions and member properties as shown in Figure A-1.

Figure A-1	Dimensions of Example Bridges
A.3.1	Applicability of Standards
A.3.2	Acceleration Coefficients
A.3.3	Importance Classification
A.3.4	Seismic Performance Category
A.3.5	Site Effects
A.3.6	Response Modification Factor
A.4.1	General
A.4.2	Analysis Procedure
A.4.3	Determination of Elastic Forces and Displacements
A.4.4	Combination of Orthogonal Seismic Forces
A.5.3	Single Mode Spectral Analysis Method—Procedure I
A.4.8	Design Forces for Seismic Performance Categories C and D
A.8.4	Seismic Performance Categories C and D
A.4.9	Design Displacements

Supplement B

Summary of Redesigns

B1	Introduction
B2	Changes to the Third Draft of the Standards Resulting from the Redesigns
B3	Summary of Cost Impact
B4	Detailed Results of Redesigns

REFERENCES

1. American Association of State Highway and Transportation Officials, *Standard Specifications for Highway Bridges*, 15th ed., 1992 and *Interim Specifications*, 1992.

REFERENCES

2. *Design of Highway Bridges*, National Standard of Canada, CAN/CSA-S6-88.
3. Ontario Highway Bridge Design Code, 1973, volumes 1 and 2, Ontario Ministry of Transportation and Communication.
4. ANST/AASHTO/AWs Bridge Welding Code D1.588, 1988.
5. *General Specification for Welding of Steel Structures*, CSA Standard W59.1-1970, Toronto, 1970.
6. *Reinforced Concrete Design Handbook Working Stress Method SP-3*, 3rd ed., American Concrete Institute, 1980.
7. *Notes on Load Factor Design for Reinforced Concrete Bridge Structures with Design Applications*, Portland Cement Association, Chicago, IL.
8. *Post-Tensioning Manual*, Post-Tensioning Institute, Glenview, IL., 1978.
9. *Post-Tensioning Manual*, Post-Tensioning Institute, Glenview, IL., 1976.
10. American Association of State Highway and Transportation Officials, *Guide Specifications for Design and Construction of Segmental Concrete Bridges*, 1989.

CHAPTER 12

METHODOLOGICAL TRENDS IN DESIGN OF BRIDGES

12.1 CHARACTERISTICS OF BASIC TRENDS IN THE DESIGN OF BRIDGES

The design of bridges, according to some nonspecialists, is based on exact analysis and for this reason is analogous to the solution of mathematical problems, where results are obtained by examining the problem data and utilizing mathematical methods to arrive at a solution. Although this view is accepted by some specialists, it does not correspond to reality. Technical and economic analyses are very important aspects of bridge design, but they do not cover the whole design process.

First, many problems cannot be solved numerically. Second, the analysis may not correspond exactly to the actual situation, therefore the results may be disputed. Technical analysis has value during construction, but not during the solution of basic problems: choice of bridge system, material, general dimensions, foundations, and so on. These problems are solved on the basis of general considerations and the designer's particular point of view.

The practice of having technical discussions to compare bridge projects usually results in differences of opinion among different designers considering the same problems. The same result occurs during discussions on the basic problems of bridge engineering.[1] The final choice of alternative depends to some extent on the composition of the participants. Therefore, it is necessary to analyze the different points of view and determine to what extent one or another is correct. Such an analysis of separate opinions should be done systematically, by establishing the general principles on which the analysis is based. Previously such discussions did not take place, resulting in problems. However, by correctly assessing the different views, prejudices and old ideas are eliminated, and the proper solution becomes apparent.

The experience of holding technical discussions and learning from past solutions indicates that in the area of bridge design there exists a few basic directions. The analysis of different methods of bridge design solves perplexities that baffle nonspecialists, especially when they witness differences of opinion among specialists. Such discussions also correct the belief that these differences in opinion are determined by personal motivations.

The existence of different methodological directions in bridge design is not strange, but should be accepted as inevitable, if we consider the history of bridge engineering. Progress in the techniques of bridge construction depend on scientific and technological development and general historical conditions. At each historical moment in the creations of designers traditions of the old are preserved and more modern views are born.

An investigation of the history of bridges indicates that the technique of bridge construction had a few stages, connected to the development of industrial methods.[2] We may call these stages primitive, industrial, architectural, and engineering. In the last stage it is also possible to consider other separate stages of development.

The influence on bridge design of older principles indicates that to best understand the present, it is valuable to study the history of bridges, which is connected to the general development of the material and spiritual culture of the society. In the transfer of past heritage and the preservation of technological advancement, the university has great influence, where the future engineer takes from his professor basic knowledge, corresponding to the spirit of the time.

12.1.1 Rational Computation Trend

In the second half of the nineteenth century the general trend of design may be characterized as rational computation. It resulted from the introduction of specifications regarding loading and permissible stresses, the development of typical designs, the separation of design from investigation and building, and the introduction of steel trusses. As a result, the process of design took on the character of office work, theoretical-analytical and drawing work, and was removed from the real conditions of bridge construction. The first books written about bridges, which were based on construction practices, hastened the acceptance of this method, which became the basis for the rational computation methodological trend since bridge theory was now considered to be a matter of structural mechanics.

The study and systematization of material gave to the new discipline of "bridges" the character of an empirical science, and structural mechanics was its theoretical part. It was impossible to provide in the books all the practical material available, and for this reason separate examples were introduced, resulting in the creation of a system of ready-made classical examples for different cases. If it were necessary to solve a new problem, examples from the past were considered. Supporters of the rational computation trend believed that bridges should have certain definite shapes.

Considering the most often used systems, a few alternatives were described. For each one, the optimum geometrical dimensions, beneficial span, height of the truss or rise of the arch, and so on, are determined and alternatives are compared. The choice of alternative is based on numerical indexes, and for this reason supporters of the rational computation trend introduced these indexes in all parts of bridge design, sometimes when they cannot properly be applied.

12.1.2 Creative Trend

The rational computation trend of bridge design was impossible to apply in the design of reinforced-concrete bridges during the twentieth century. New systems and constructions appeared and changed very quickly. Nothing was repeated; new rules were always provided. Also, the properties of new materials were not investigated. Each design analysis was carried out individually. There were no technical rules; everything was left to the designer. It was necessary to create new specifications on the basis of general knowledge regarding the performance of reinforced concrete structures. Old methods were actually obstacles.

The building of highway bridges and the application of reinforced concrete put before designers in every case a basic problem regarding the choice of bridge system, and for this reason preliminary design became important. The concept of reinforced concrete bridges forces the designer to consider the bridge not as a summary of separate parts, but as a monolithic unit, where changing one part leads to the changes of other parts. Here the strength is not the basis for design. It is necessary to design the structure and then to analyze it.

At the beginning of the twentieth century, owing to the introduction of reinforced concrete and the development of suspension bridges, the designer had a large choice of materials and bridge systems. It was possible to develop similar rational alternatives.

The conviction that in each particular case there exists not only one rational solution, but a few from which it is possible to choose according to one's own taste, represents the basic characteristic of the second trend in bridge design, which may be called "creative."

The design of each bridge is the solution to a new problem. There does not exist a ready solution, it must be looked for. Considering the role of personal creation, this second trend may provide original, new projects. Supporters of this second trend believe that the creation of a bridge depends on the personal taste, capability, and imagination of the designer. Design is considered as the creative process consisting of the combination of structural parts through the use of knowledge gained from the study.

First, the designer should compose the scheme of the bridge. This work is based on the imagination. The designer prepares a number of possible alternatives for the general shape. The second stage follows—preliminary design. The chosen scheme undergoes detailed development for further comparison. Here the main process is self-investigation and critical analysis.

The preliminary designs are developed, after they are compared and the final alternative chosen. The third stage follows—detailed design, in which the many parts and details are chosen by existing analytical and construction methods. However, during the development of details, problems arise that require investigation and rules and specifications may be disputed, because they are not firm dogmas.

12.1.3 Practical Trend

The first consideration in this new trend is practicality. And such practical design requires scientific investigation. Practical designers use both scientific principles and creativity in their designs, but only insofar as they solve practical problems. They consider that the bridge is part of the highway or railway and its basic purpose is to satisfy the requirements of transport.

The bridge should possess the best exploitation qualities; the basic requirements of the bridge are economical; the construction of the bridge should be adjusted to the industrial fabrication.

Supporters of the creative trend considered highways and railways as areas to apply their creative capabilities, testing their new inventions. Scientific investigation workers considered highways as a large laboratory for their investigations. Practical designers were against this, asking that the bridges be safe, testing structures, and permitting experimental structures only on secondary highways. Practical designers suggested that structures be standardized for industrial preparation. In the case of bridge reconstruction, practical designers suggested the fastest reconstruction. Also practical designers suggested the construction of structures that required minimum maintenance, which would not affect traffic.

12.2 BASIC ASSUMPTIONS OF DESIGN

The modern realistic method of the design and construction of bridges is based on technical expediency, economy of construction, satisfying the requirements of convenient planning, and introducing progressive methods of bridge design.

Aesthetically, the external shape and type of structure should be expressed by the simplicity and austerity of the shape, economy in structural solution, exact proportions, and expedient use of materials. The designer should develop methodological rules that correspond to the technical and practical requirements of bridge engineering.[3-5]

Basic problems should be solved in the process of preliminary design. The first requirement of a bridge, as for every structure, is expedience. To indicate the real contents of this general requirement, it is necessary to define more accurately the aim of the structure and therefore the design of the bridge. It is known that the bridge is designed to cross rivers or other obstructions and to satisfy the requirements of transportation. This basic designation or aim of bridge building was sometimes complemented by the solution of other secondary problems:

1. Roman bridges and those built during the Middle Ages served as places for walking and were even used for dances. These traditions were continued to later times.
2. Sometimes bridges were considered as monuments.
3. During the sixteenth and seventeenth centuries bridges served for commerce and were built as wide structures. Most of their area was used for shops. This use was dropped toward the beginning of the nineteenth century.
4. During the Middle Ages bridges sometimes were considered as fortresses and had towers. Later, such towers or massive stone portals were built only for architectural considerations.
5. During the same period chapels often were installed on the bridges and tolls were collected for use of the bridge.
6. During the Middle Ages and somewhat later bridges were partially used as dams for the watermills.

The basic purpose of the bridge—to serve the requirements of transport—is most clearly shown by the railway and automobile highways built between cities. Transport requirements consist of the safe and convenient traverse over the bridge. The requirement of safety is imperative and is assured by the carrying capacity of the bridge, or the maximum value of the temporary vertical load that the bridge can take. No interruptions in traffic can be assured by calculating the maximum number of vehicles passing in a specific unit of time. For bridges across shipping rivers and also for underpasses, we must consider passing capacity under the bridge. Passing capacity of the bridge is characterized by the number of lanes, their width, and accepted lateral clear distances of shoulders and medians required for safety considerations.

Passing capacity of the bridge is calculated by the capacity of the highway. Sometimes to assure uninterrupted traffic it is necessary that the width of the bridge be greater than that required by the calculated passing capacity. For example, on long bridges it is necessary to foresee possible cases of forced parking, which may create an interruption in the traffic. As a rule, the width of the roadway on the bridge is equal to the width of the highway. However, there may be deviations from this rule. For example, although the highway may have three lanes of traffic, on the bridge the number of lanes may be reduced and the cost of the bridge substantially reduced. On the other hand, for small bridges, the width of the roadway on the bridge may be greater than the width of the highway.

Difficulties with the traffic may also result from the grades at the approaches to the bridge, from wear and tear, and frequently from repairs being made to the roadway surface. These difficulties define the degree of traffic comfort.

Maximum traffic comfort is not a requirement, but it is preferable, and this would be one of the criteria appraised during planning of the project.

The design of the structure is not limited to the necessity to satisfy the required conditions; the designer should consider economic conditions—the necessity of minimizing the labor and time needed to build the bridge. Here we consider not only labor, which is necessary for the construction, but preparation of the material. The amount of labor and material is estimated from their cost, the quality by the time available for construction.

Therefore, the most expedient bridge is the structure that best satisfies the requirements of transport, passing capacity, and most traffic comfort conditions, has the minimum cost, and may be erected in the shortest time. All of these characteristics are connected. One characteristic of modern transport is an increase in the weight of vehicles and their number, which requires that the carrying and passing capacity of the bridge be designed to allow for more weight. The greater the reserve, the longer the bridge may exist without changes to or reinforcement of the construction. However, increasing the reserve of passing and carrying capacity increases the cost of the bridge. Therefore, the problem of determining the necessary reserve of passing and carrying capacity should be solved from an economical point of view.

From this comes the negative concept of an "endless" bridge: In this case it is necessary to give to the bridge infinitely great reserves of passing and carrying capacity, resulting in an excessive increase of bridge cost. The origin of the idea of endless bridges has a historical basis. The Romans built massive stone bridges instead of using timber, which was subject to wear and decay, and by comparison with timber, stone bridges may be considered as endless: the temporary load on the bridges of people or horse vehicles was very small by comparison with the weight of the structure and may be considered as unchangeable. Romans did not visualize the development of transport, and for this reason the idea of an endless existence of the bridge was natural.

Developed at the end of the nineteenth century, this idea was in basic contradiction with practice, because it neglected the idea of development. It is necessary to note that in many cases the expression "endless bridge" is only figurative. The service life of an endless bridge is often considered indefinite, but is actually defined by the properties of the material used. Such a definition of the service life of the bridge cannot be thought of as a designation its of carrying capacity reserve and passing capacity. First there is no definite method to establish the service life; second, it leads toward attempting to establish an increased reserve "for every purpose," because usually a bridge's existence is assumed to be 60–100 years.

The property of the material is not necessarily the basic factor that defines the service time of the bridge. Generally bridges are reconstructed for other reasons, for example, owing to small passing and carrying capacity, insufficient clearance under the bridge, straightening of lanes, or reduction of the grade. History provides a great number of examples of such reconstruction.

The necessary reserve of carrying and passing capacity, and thus the service time of the bridge, are determined by economic considerations. The idea of greater service time of the bridge should be rejected when determining the reserve of carrying and passing capacity, but accepted as guidance during construction of the bridge. This requires detailed and careful structural work from the designer such that the construction itself and the material used will not wear out before their time. If, due to bad construction and the destruction of materials, the bridge is not able to serve for the time defined by its carrying and passing capacity, then its reserve will not be used and the expenses involved will not be justified. If it has good construction, the bridge may outlive its economical expedience time, and may be used to carry lighter loads or as the basis for a new or reinforced bridge.

Normally the carrying and passing capacity of the bridge as calculated loads and clearances are chosen by the designer earlier and serve as a premise for the design. The cost of the bridge and time for construction are determined by the project.

12.2.1 Basic Requirements of the Bridge Under Design

Starting with the established premises, the basic requirements of the designed structure are determined, which, during appraisal of the project and comparison of the alternatives, transfer into the criteria for the appraisal.

Safety conditions require that the bridge be convenient, economical, and timely in construction. Technical conditions require that the bridge as a whole and each element of it separately be durable, rigid, and stable. This is checked by the analysis using accepted specifications. But not all questions of durability, rigidity, and stability may be covered by the analysis—in some cases it is necessary to provide construction measures that consider the performance of the structure under loading.

Specifications and technical conditions should be satisfied because they guarantee the carrying capacity of the structure. From the point of view of safety all bridges designed according to the technical requirements are equal. However, practically speaking, technical requirements may be satisfied in different degrees. In some cases they are accomplished exactly, in others with reservations. Special meanings are given to the technical criteria for such parts of the bridge as foundations when they are designed; however, degrees of safety may differ and there are often doubts about their safety.

Regarding different elements of the bridge it is necessary to know that for the engineering structure the best example may be the body of equal resistance.

Therefore, it is reasonable to have equal reserves of carrying capacity at all elements of the structure. Generally, during comparison of projects, technical requirements should be considered. But because technical requirements may be accomplished using any alternative, consideration should be given to additional guarantees of safety.

Exploitation requirements naturally should have great importance, but basically they are satisfied by accepted clearances; therefore during design it is necessary to consider greater or smaller convenience of the traffic. First, the height of the bridge or elevation of the roadway must be determined, followed by the greater or smaller grades of the approaches. Maximum grades are according to the specifications, but it is convenient to have the minimal grades. Further it is important to define the number of joints in the roadway, which depends on the division of the structure into separate sections.

The requirements for the minimal wear of the parts of carrying construction under the influence of moving vehicles are necessary to consider. Regarding the maintenance of the roadway and the bridge, it is possible to consider this as a general cost, and, therefore, relate it to the economic requirements.

Exploitation requirements indicate that the cost of the bridge at all equal conditions should be minimal. The general cost of the bridge is determined by the quantity of material and unit price, which depends on the cost of building and erecting of the structure. For this reason the tendency to reduce the quantity of material does not always lead to the minimal cost. For example, sometimes steel structures are designed without considering convenience of their fabrication at the plant, paying attention only to obtaining a minimum of material. However, advantages may be gained during industrial preparation that result in convenience of erection. This may result in a heavier construction, but the general cost is smaller.

During comparisons of various projects, an analysis of their economic criteria may reveal principles of expedience that can be applied to the project under consideration. Building requirements are connected to economic constraints, because when the amount of material is small, the work is simpler and the time required is shorter. Also, the unit price is considered as part of the economic criteria, indicating the cost of preparation and erection.

For conventional bridges built from the same material, construction is carried out by established methods; therefore, during comparison of alternatives, purely construction criteria are not important. In special cases of complicated erection of bridges having large spans, or for urgent work, building requirements are very important and may influence the choice of bridge system and material. These requirements also obtain special importance from the point of view of the introduction of mechanization. In these cases it may be necessary to use a great quantity of materials, increasing the cost of construction, and ignoring other requirements. For example, during the initial period of application, assembled reinforced concrete constructions were more expensive than monolithic ones; however with increased use of these constructions, the application of assembled structures becomes more rational and economical.

12.2.2 Additional Requirements of the Bridge Under Design

Aesthetical Requirements. Apart from the basic requirements of the bridge under design, there are often additional requirements, the first of which is the problem of aesthetics. Beauty should be achieved as a result of the good proportions of the whole bridge and its separate parts. In spite of the tendency to build economical structures, we should not renounce beauty. The architecture of the bridge should not be ignored because of economic and technical requirements.

There are different views regarding the aesthetical requirements in bridge engineering. Supporters of the rational analytical trend feel that aesthetical requirements are not important and not necessary for bridges outside cities. Designers of the creative trend consider these requirements to be more important than economical ones and equivalent to the requirements of strength and longevity.

Because of such conflicting views, this problem requires special consideration. Every designer inevitably tends to want his structure to be the most beautiful. This wish is natural, it shows love and interest in the work, and it transfers into the tendency toward perfection of the designed structure.

During the process of design the engineer is occupied with detailed calculation and construction, and may be occupied with particularities, deviating from the complete structure. By checking his creation from an aesthetical point of view he gives his attention to the general concept and shape of the structure. This gives him a better opportunity to design details and to correct if necessary the general scheme of the structure. If the designer is aesthetically unsatisfied with his creation, this forces him to improve it and finally to find a convenient solution. But all this is possible only when the designer is not following any preconceived aesthetical rules and is not going outside the limits of technical and economical expedience of the structure. In relation to the architecture of bridges this means that the external view of the bridge should not contradict either as a whole or in its details the purpose of the structure and its technical concept.

Technical literacy is necessary; however, sometimes it is not enough to obtain a good external view of the bridge. Technically and properly designed bridges in every case achieve beauty. However, if the designer is guided by the caprices of his own taste, regardless of the technical truth, he will not achieve his goal. A beautiful shape cannot be invested and applied to the bridge. It should result from its technical concepts, from its structural shape.

The critical rules of proportion and the use of purely geometrical shapes had, in their time, not so much an aesthetical but a technical production basis. They based their theories on the principle of imitation of relations, which they observed in nature. Historical investigation indicates that many contemporary aesthetical rules are preserved from previous centuries, when they had a different basis. Even today a bridge is considered beautiful when it has an even number of supports. According to Palladio, it is clear that this rule was accepted because all birds and animals have an even number of legs, which give them better stability.

Freeing themselves from prejudice, and carrying out independent investigations to find the shape corresponding to the contents, should lead designers toward the development of a theory of true aesthetics in bridge engineering. Historical description shows how the shapes of bridges were changed depending on the general development of the culture and economical life of a nation. For this reason the problems of aesthetics in bridge engineering should be viewed in a historical perspective. The designer should be able to determine the time when the bridge was built by considering an external view of the bridge, its scheme, and construction.

Followers of the historic direction renounced such investigations and by this changed their principles and attracted architects to the design of bridges. However, a joint venture by designers and architects is not always useful for solving a problem of bridge design. Architects specialize in the construction of buildings, which is reflected in their aesthetical taste. Although architectural rules and views may be correct for building, they may not apply to the architecture of bridges. For example, when designing a building, architects usually use steel construction as a frame for the building, which requires a certain covering. For the bridge designer steel construction is a force polygon that clearly demonstrates the transfer of forces. For an attractive external view of the bridge, detailed design and proper accomplishment of the construction are important. The external view may be spoiled by careless work.

During correct design of the bridge the technical concepts of structure and architectural shape should be not separated and are shown from the local conditions and general requirements. By underlining the validity of the aesthetical requirements it is not necessary to recognize special aesthetical criteria for choosing an alternative. The final choice of the alternative is the solution of some technical problem in correspondence with the basic purpose of the bridge as part of the highway.

If the bridge is not considered as a monument but serves only for traffic for a certain period of time, then it is unnecessary to request from the designer that the bridge represent a high artistic creation. We may be satisfied by more modest wishes as regards its external view. Practice indicates that designers may create, and actually have created, attractive bridges even when they were governed only by the technical and economical requirements during the design process.

The bridge that is correctly designed from the point of view of technical-economical criteria cannot contradict the basic rules of architecture. The general basis of architecture valid for all structures consists of the idea that the masses of material should be distributed expediently, the properties of the material should be used correspondingly, and the whole structure should correspond to its purpose.

Generally, economical considerations during bridge design actually are the same as stated above. An economical design is achieved first by the expedient distribution of material (choice of most economical system, change of the cross-sections of the members, considering conditions of their work); second, by the use of proper material (members in tension use steel; members in comparison use concrete; the sections are chosen with small reserves of their strength, etc.).

Therefore, economical expediency and architectural conception are determined by the same criteria. From this it follows that is impossible to contrast the aesthetical criteria with the technical-economical. For example, it is advisable to reject a beam bridge for an arch in the case when the first by all other properties is better, or to prefer a single-span bridge to the more expedient two spans. Also, it is possible to say that the choice of alternative, considering technical and economical criteria, shall not deviate from the proper way to achieve also aesthetical aims.

Finally, the bridge will only be perfect in an aesthetical sense, when its system as a whole and its separate members are chosen not on the basis of personal taste of the designer, but considering technical-economical expedience.

All other proofs, which are often applied by the authors of separate projects to defend unsuccessful technical-economical alternatives, should be rejected. All these proofs are based on the unstable and changeable bases of personal taste; such proofs are only declarations of personal impressions and tend not to prove anything but only to convince people by the use of verbal arguments.

Many definitions are expressed using varied terminology synonymous in meaning but with drastically different shades in the positive and negative sense. For example, regarding structure of the bridge, when the deck is at the bottom chord the defender may say that this structure is "expressive," "easily seen," or "stands out with a beautiful shape on the sky," but the opponent may object that this structure "obstructs view," "hangs on the observer," and so on. By the skillful use of such terminology it is possible to convince the inexperienced that a beautifully presented perspective is not as worthy of praise as a less successful project.

Requirement of Scientific Research. The second additional requirement sometimes asked of bridges under design may be called the scientific research or "innovatory." This requires that the bridge contain a new achievement due to scientific research or a new invention.

The design of every bridge always contains something new. Even if the project is worked using old examples and applying typical projects, the designer uses new combinations—the known under new conditions. Therefore, there is always a certain degree of novelty. In this connection, the following may be requested from the designer: He should not only be familiar with the old designs, but should know modern scientific research and use it in the design, depending on the project. Considering such requests to improve the project, he may carry out his own investigations and offer his own rational and inventive proposals.

It is natural for the designer to search for novelty, but new solutions should be born only from the contents, from the tendency to reach the best, starting from the real conditions of the project. Therefore the "novelty" requirement cannot be opposed to basics, because both should be directed toward one aim.

From the history of bridge design it is known that the development of bridge engineering generally, and considering separate types of bridges, did not proceed uniformly. Some periods were distinguished by invention and the appearance of new shapes and systems of bridges; other periods were characterized by the mastering and improvement of existing systems and the development of scientific research work. For example, at the end of the eighteenth and beginning of the nineteenth century, a great step forward was made in the area of stone bridges. At the same time different timber trusses, of large spans, were developed and the cast iron arch and iron suspension systems appeared. All of these novelties resulted from the development of transport and industry. Forty or 50 years of the last century were spent creating the iron beam bridges, and the second part of the last century was devoted to the development of expedient systems and the improvement of construction. Similar periods in later history were devoted to the development of reinforced concrete bridges. The initial period of trials and creation of the construction was 1880–1890, and the period of mastery was 1900–1910. At different times scientific research work changed the role of novelty in the creation of the structure and its improvement. However, it is necessary to note that with the general development of science and technology, the role of scientific research is increasing together with novelty, rationalization, and invention. From this it follows that the necessity for novelty results from the general economical conditions and requirements.

The attitude toward novelty in bridge engineering was different and served as a subject of discussion. Adherents of the rational-analytical direction preferred generally to hold to some classical models, considering that the search for new shapes should be related only to scientific research work. Adepts of creative direction, however, were tending toward the new and original, by omitting any old pattern. Realistic understanding of the new should be based on an understanding that novelty is not an aim in itself and that the new should be better than the old, otherwise it has no sense.

From this point of view it is necessary to consider the criterion of novelty because it sometimes appears as an independent factor during appraisal of projects and choice of alternatives. Because the novelty is not an aim in itself, it should not be a special criterion, forcing a preference for new construction irrespective of its quality. From the appraisal point of view it should be placed on an equal footing with the old. When by basic conditions the new is better and there is no doubt regarding its quality, then it should be preferred to the old, and in the opposite case should be refused.

Not every novelty leads toward progress in bridge engineering; some novelties go against proper development. If the novelty is sound and capable for the development, it may be developed to such a degree that its quality is better than the old. But if the novelty is not better than the old one or not yet developed, it is necessary to help its development and abstain from early application. Early application leads toward lowering the quality of bridges and may compromise new ideas.

The criterion of novelty may be considered independent only in separate cases when economy requires the introduction of new constructions. An example is the introduction of prefabricated reinforced concrete constructions. At the present time it is expedient to use prefabricated reinforced concrete constructions, but initially they were more expensive than conventional constructions. The criterion of novelty then was in controversy with other criteria. This controversy had to be solved in each particular case, especially when the novelty was not an aim in itself, but was required for economy.

One reason for introducing new construction is related to the necessity of experimental and practical checking of the results of scientific research work, which is certainly necessary. Regarding bridges on main highways, however, it is not advisable to subject them to experiments, because their basic designation is to serve transportion in the country. Only separate experimental structures and special controls are permitted. However, in each case, the problems of special scientific experiments and experimental structure should be solved at a scientific institution.

12.3 BASIC PARAMETERS OF THE BRIDGE

The quality of the structure is estimated considering different criteria: technical, exploitation, economical, construction, and additional, depending on the material system, and geometrical dimensions of the bridge. All these criteria are temporary parameters defining the quality of the structure.

The problem of design generally consists of the way to find the values of these parameters that will correspond to a better quality of the structure. For this it is necessary to consider first, in detail, basic factors influencing the quality. All the parameters are connected together, but their influence on the quality of the structure is different. Reciprocal connections of the parameters also differ: one may depend little on another, yet may greatly influence the other.

During preliminary design toward basic parameters the following are also related: location of the bridge, span, material, type of foundation, system of the bridge, length of separate spans, type of span construction, and type of supports. Clearances and design load are not included here, because they were discussed earlier.

Location of the bridge usually does not depend, or depends little, on the other parameters, but does influence them. For small bridges the location is defined by the intersection of the highway with the river, ravine, and so on. For medium and especially large bridges it is possible to compare a number of location alternatives, such as the basic value of the highway, the cost of approaches, and highway installations. The cost of the bridge itself has no de-

12.3 BASIC PARAMETERS OF THE BRIDGE

ciding role because its span at all alternatives is usually unchangeable. For this reason during selection of bridge location it is possible to propose an often-used bridge type without detailed study. However, there are two exceptions to this general rule: If the river is not for shipping, and has sandbanks, then at the location of largest curvature the span of the bridge obtained is smaller, but the depth of the water here is greater. Therefore, foundations are complicated and the installation of pile supports may be impossible. On the sandbanks where the span is increased, but the water depth is more shallow, it is possible to build a simple viaduct-type bridge supported by the piles. If the bridge is proposed to be built from timber, then its location should be chosen over sandbank. Therefore, during choice of crossing it is necessary to consider in both alternatives different types of bridges.

The second exception is the design of viaducts across mountain ravines. Here change of the crossing substantially changes the span of the bridge and correspondingly its cost. It is true that the type of the bridge for the first comparison may be left the same (e.g., reinforced concrete arch type), but it may be designed for all alternatives, because the cost of the viaduct will influence the location of the crossing.

The indicated exceptions do not occur often and should be considered separately; for this reason the location of the crossing may be chosen before preliminary design and must be made by the investigators with designers only checking the correctness of the choice. The size of the bridge opening is defined by the hydrological and hydrometric investigations and during design is used as assumed. However, in some cases, during the design process it is possible to change the span. The size of opening, as shown above, depends on the crossing location. It depends also on the type and depth of the foundation: at greater depths it is permitted greater wash-out, with corresponding diminishing of the opening, and at shallow foundations the reverse. In principle two opposite solutions may exist: to build bridge supports as safe against washing, squeeze the river by flow-directed dikes, and obtain the opening as minimal, or not squeeze the river, cross the whole river during flood, and thus the concern that the supports will wash out.

The first solution is used as a rule for rivers on the plane. It may be justified economically and technically. Only at timber bridges is it expedient to cover the whole flood area or the great part of it by the approach viaducts. Here the size of the opening depends on the bridge material. The second solution may often be expedient for mountain rivers in which the main channel is often changing and threatens to wash out the flood embankment.

Considering these special cases it is generally possible to establish that the size of the opening is provided and during the process of design may change a little depending on the type of foundation, and if the type of foundation as a whole is determined by the local conditions (e.g., by using caissons or wells), then the size of opening for all alternatives remains unchanged.

Bridge material is a basic parameter, affecting the whole structure. If the material of the structure is not chosen, it is impossible to investigate other parameters of the structure. Choice of material is the most substantial problem during preliminary design, but it depends not only on the designer, but on the other conditions and must be considered before preparation of the project. Every material has its own area of application, and the problem of material choice arises when these areas cross each other.

Timber bridges are usually considered as temporary structures and design of spans greater than 160 ft present difficulties. For permanent bridges the choice is between reinforced concrete and steel structures. For spans from 65 to 100 ft, reinforced-concrete beam-type bridges are mainly used and steel is considered for overpasses and underpasses. At spans greater than 330–500 ft the preference is for steel bridges, and for those greater than 650–800 ft it is expedient to use steel bridges. Therefore, the choice between reinforced concrete and steel is generally at spans of 65 to 330 ft.

The type of foundation is determined mainly by the geological investigation, and in the main channel of the river also by the depth of the water. This parameter depends comparatively less on others (dependence of the bridge location and size of the opening are indicated above), but the type of foundation basically influences the superstructure, sizes of separate spans, and type of supports. Foundations built at the present time may be divided into two basic groups:

1. Timber piles, which are used for shallow foundations, and reinforced concrete and steel piles, which are used for deep foundation.

2. Massive, shallow foundation (between others or piles) and deep foundations (caissons and wells). At costly foundations the spans should be greater. It is obvious that for the large spans it is necessary to use massive foundations. Shallow pile foundations are possible for viaduct bridges having small spans. Regarding the bridge system, the following considerations apply: pile foundations almost define the beam system, arches and suspension bridges require massive foundations and massive supports, but there may be other alternatives. Considered parameters are of prime importance and generally are determined before the design is started and are based on investigations. They depend comparatively less on subsequent parameters, but to a great degree influence them. During design the following four parameters remain constant or change little.

12.3.1 Bridge System

The system (beam, arch, suspension) is substantially connected with the material. Beam systems are mostly used; arches are mainly for large spans, and suspensions for great spans.

For reinforced concrete bridges the following conditions apply: at spans up to 130 ft—mainly beam systems; at spans 130–200 ft—beam and arch; at greater spans—arch. For steel bridges beam systems are mainly used. The arch system is expedient at spans of 160 ft and greater.

All above data are approximate and are given only as a starting point. The bridge system depends on its other parameters and establishes connections among these parameters. It is impossible to investigate the other parameters without assuming the material of the structure and the bridge system.

12.3.2 Size of Separate Spans

This factor substantially influences the cost of the bridge. Therefore determination of the span size enters into a number of basic problems solved during the preliminary design.

For beam bridges having steel trusses a known rule exists, obtained mathematically: The cost of the main truss with bracing per one span should be equal to the cost of one pier with foundation. For all other cases it is possible to say that with the increased cost of the pier and foundation, the size of the span should be increased, or the size of the span depends on the type of foundation and pier.

Similarly, the system of the span construction has influence on the size of the span. With arch bridges the cost of the support is generally greater than with the beam type. For this reason, all other conditions being equal, the span of the arch bridge should be greater than that for the beam type. The exceptions are high viaducts having rising arches. For such arches, a comparison with low arches having the same size span indicates a more economical cost. The limits of changes to span sizes are governed by clearances for ships and typical uses of the span structures. The clearance for ships regulate the minimal size of span. Usually this span is greater than the most economical length. For this reason during crossings of navigation rivers the size of the main span at the main channel in most cases is predetermined. It is necessary to change only side and approach spans. During the choice of approach spans it is necessary to consider typical projects, because the use of typical constructions is more rational and useful.

From this it follows that the sizes of spans are not arbitrary—they are chosen from defined sizes. The span size is closely connected with the system of span structure; therefore, it is necessary at the start to assume a system of span structure. However, span size often determines the bridge system.

12.3.3 Type of Span Construction

Span construction is closely connected to the bridge system. After assuming a bridge system, span structure is assumed. There may be problems regarding the type of structure (e.g., solid or truss type for steel, monolith or prefabricated for reinforced concrete), the number of main girders, the basic dimensions, and so on.

Therefore, this general parameter includes a great number of particular parameters that should be determined. Detailed study of them in each case requires quite a bit of effort and time. But many problems common to particular cases may be investigated earlier, during the preparation of typical projects.

The use of typical projects substantially helps individual design. For the majority of medium-span bridges, typical projects may be used. The use of typical projects simplifies fabrication of the structure, reduces the time necessary for design and construction, and makes the structure more economical. However, the immediate use of typical projects should not be considered as a rule. They should be considered as a first solution, which in many cases can be improved. It is always possible to design a new structure which corresponds to the given conditions. Typical projects do not take into account all possible cases of design, for example, design of large-span bridges. Also, examples of already built bridges, may provide a rational starting point. Together with this experience in the design and building of bridges permits us to establish some useful relations, for example, the ratio of truss height toward span, to the number and length of panels, and so on.

The design of bridge structures starts with the critical study and use of existing bridges to prepare the first alternative of the structure, and continues during the investigation of separate parameters to prepare the next alternatives.

12.3.4 Types of Supports

Supports may be divided into two groups: columns and massive supports. The second group is used in the presence of large floating ice and the arch-type span structures. Column-type supports are most expedient with small-span beam structures.

12.4 THEORETICAL BASIC METHODS OF PRELIMINARY DESIGN

12.4.1 Introduction

Methods of design cannot be invented on the basis of certain arbitrary principles; they are created by the practice. In a given historical study we have enough proof that the methods of design were changed depending on the bridge building practice and its basic problems. At the present time, the method of preliminary design applied also is determined by practice. There is no reason to reject it and substitute some new one.

The problems consist of the exact application of contemporary method, the explanation of its logical basis, the removal of some controversies and, by this, their improvement. Considering modern conditions, the problem of preliminary design consists of creating a bridge of the best quality from the point of view of established criteria. The quality of the bridge depends on the parameters considered.

Using mathematical designations, it is possible to write (Fig. 12.1) that the quality index U of the structure is a function of its parameters, x, y, z, or

$$U = u(x, y, z \ldots) \qquad (12.1)$$

During preliminary design values of parameters should be determined, at which the bridge may have the best quality indexes. This problem is similar to the problem of variation calculus to find the limit of the function. This analogy may be used partially to determine a logical basis for the method of preliminary design. It is clear that the problem of preliminary design cannot be solved purely mathematically, because the quality indexes cannot be expressed as any algebraic function, and the majority of parameters from one

12.4 THEORETICAL BASIC METHODS OF PRELIMINARY DESIGN 215

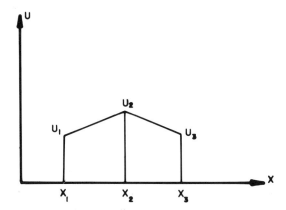

Figure 12.1 Quality index of the structure.

alternative toward another change by quality, but not by quantity, and sometimes at a rapid rate.

Only in particular cases may a mathematical method be applied to find the limit. For example, it is known that by this method were found the sizes of the most beneficial spans of simple-span trusses, the most beneficial heights of steel trusses having parallel chords from the condition of minimal weight, and so on. But if it is impossible to apply mathematical methods directly at a composition and solve equations, then its logical basis and reasonings may be applied as a scientific method for the solution of each problem.

To find the limit of function U it is possible to find corresponding values of parameters x, y, z from the equations

$$\frac{\partial U}{\partial x} = 0; \quad \frac{\partial U}{\partial y} = 0; \quad \frac{\partial U}{\partial z} = 0 \quad (12.2)$$

The sense of these equations is in the investigation of the influence of each parameter in the change of quality indexes of the structure.

By leaving all other parameters constant, we give the increase (generally speaking, changing) $\pm \Delta x$ to the given parameter and study the change of the value ΔU. By this method we find the value of the parameter at which the increased ΔU changes its sign. This corresponds to (limit) of function U, depending on the first parameter x.

It is necessary to note that the separate parameters are not in this case independent one from the other, and because of this, sometimes by changing one parameter it is necessary to change another. For example, by increasing over a certain limit the span of the reinforced concrete bridge, it is necessary to move from a beam system toward an arched. By application of this method of preliminary design of bridges the composition of (12.2) leads to the composition of alternative of crossing. For each equation of (12.2) it is necessary to use a minimum of three alternatives. The first alternative is composed of certain values of the parameters x_1, y_1, z_1 . . . . . After leaving parameters y_1, z_1

... constant, we give to parameter x a new value x_2 and compose the second alternative. By comparing it with the first, we establish the change of quality indexes of the bridge. If they improved, it is necessary to change once more parameter x in the same direction by giving to it the value x_3, and compose a third alternative. By comparing it with the first two alternatives we determine the change of quality indexes of the designed bridge. If, for example, they become worse, then their maximum corresponds to the value x_2 (see Fig. 12.1). But if they are improved, then it is necessary to repeat the investigation (composition of alternatives) for composition of the second equation $\partial U/\partial y = 0$, further for the third, and so on. It is necessary to solve all these equations together. Regarding the preliminary design, this means that it is necessary to compose a great number of alternatives and then to consider them together. This is a very difficult and tedious process. The difficulty is increased because, unlike the purely mathematical method, where the function U is given, during preliminary design even type of function is unknown.

On the basis of these difficulties the statement by the supporters of the creative direction may be explained, that in the first stage of design everything should be based on creativity or invention. How to build a bridge over a certain river? It is possible to give a number of different answers to this question. It is possible to build a steel bridge or reinforced concrete, and for each type it is possible to give a number of alternatives. If for the given problem there are hundreds of known data, then the unknowns may number in the thousands. To design a bridge is not an easy task; it is actually very complicated. Many unknowns are found by the work of creative forces. However, it is impossible to accept their belief that the difficulties indicated cannot be solved by systematical investigation because of the following:

1. Equations (12.2) were applied to the problem of structural design, which was solved by the method of successive approximation. This method is considered legitimate in mathematics. To improve and accelerate the method of successive approximation it is important to successfully choose the first approximation, which may be based on practice. The experience of building and designing bridges provides enough examples of rational solutions for different cases. It is not advisable to accept them, as do supporters of rational-calculated direction who considered them as classical examples which should be imitated. Many reasons exist to use the experience, to accept the most convenient solution as the first approximation, refusing further investigation.

2. Preliminary design generally is the first approximation in the creation of a bridge project. It solves the question of the most important parameters, having a great influence on the quality indexes of the structure. Details of the structure may be investigated later. As already indicated, many solutions developed in practice and reflected in the typical projects exist for the details of the structure. There are not many basic parameters; therefore there is only a small number of unknowns.

3. Some parameters are given ahead and remain constant during design, others may take only a limited number of values, and this shortens the number

of alternatives. Relations among parameters, their connections and importance for quality indexes of the bridges, firmly established by practice, make it easier to carry out the methods of investigation of alternatives.

By analogy with the problem of variation calculus, one more basic difference between given methods and the views of supporters of creative direction follows. If it were possible to solve the design problem purely by mathematics, then the solution would be simply to solve equations. But it is known that all equations, except equations of the first degree, have a few roots, and differential—arbitrary—constants. It follows that in the solution of the bridge problem by the design method, in every case we obtain a few solutions, which equally satisfy the conditions of the problem. Supporters of the creative direction consider that from these equally rational solutions it is possible to choose one corresponding to personal taste.

The motive of personal taste should be refused, because it is contrary to the scientific-objective method and particularly to any useful analogy. Actually, the equations have a few roots and arbitrary constants appear in the solutions of differential equations. By applying these equations toward the solution of every concrete problem, the values of arbitrary constants are not driven by taste, but are determined from the actual conditions of the problem, for example, from the boundary conditions. A general solution of the differential equation does not indicate the solution of the problem. The determination of arbitrary constants always follows. In the application to bridge design this means that if a few equally valid alternatives are obtained, then the investigation should be continued and done more accurately.

It is possible to assume that for each actual case one most rational solution exists, and this assumption is not arbitrary, but corresponds to the analogy with mathematical equations. In mathematics books the theorem of existence and the soleness solution of differential equations is proven. Pikar proved this theorem by the method of successive approximation, which is used in this case as an analogy. Independently of this, the indicated assumption should be accepted as methodological principle, because during the solution of any practical problem we should always consider that the given problem has a certain best solution. It may be impossible always to find it, but it does exist. We may stop at the approximate solution, but we have to be convinced that by comparison with other solutions it is the best approach. The designer should be convinced on the basis of investigation of the correctness of his choice. He should defend his alternative to the customer and not show him other alternatives. Alternatives are shown for consideration only and to indicate the investigation process as well as to prove the correctness of accepted alternative. During the design process and investigation of alternatives the designer should by all means tend toward objectivity even if this does not conicide with his own taste. Design has much in common with scientific investigation. The difference is in the methods of investigation, the common part is in the unity of the method.

The work of the designer during drawing of the structure and its calculation is a special type of engineering experiment. Before the technical sciences were developed, the engineer made a model, studied it, made corrections, and tried it under the load to determine the general capacity of the structure and the strength of the details. At the present time the engineer-designer, instead of constructing a model, draws the structure on paper, and instead of trial by load, analyzes the structure under a given loading and determines stresses and deformations. The model has become a drawing and the analysis was once a test under load. But because the drawing and analysis are substantially simpler and faster than preparing and testing a model, the possibilities of such experimentation substantially increase. The contemporary designer may "build" and "break" his structure on the drawing many times, trying to find the best solution. These trials—a gradual approach toward the goal by methodical investigations—are thus similar to scientific investigation.

12.4.2 Practical Methods of Preliminary Design

At the preliminary design basic parameters of the bridge may, for convenience of investigation, be divided into separate groups depending on their role in the design. The location of the bridge and its span are given and remain as constant. The material of the bridge and its system depends on other parameters. By not assuming them, it is impossible to investigate indexes of other parameters. But a chosen bridge system for a given material stipulates the general scheme of the bridge. A scheme in which all remaining parameters enter in a definite connection, is similar, by analogy, to function U.

Starting the design by using a certain scheme, it is necessary to assume function U. The remaining parameters—type of foundation and supports, construction of the span—are unknown, but should be determined: $x, y, z \ldots$ (from them the type of foundation is often given by the geological conditions). Therefore, the following sequence of the design may be considered: If the material is not given and it is necessary to choose it, then the design process is done first for the bridge from one material, for example, reinforced concrete, and repeated for another, such as steel.

The first alternative of the bridge using the scheme is composed for one material, mostly from practice, for example, a beam span. In this scheme we need certain definite values for the parameters x, y, z. For further investigation it is necessary to change one of the parameters. The question may be asked—which one? It is obvious that it is necessary to start by investigating the parameter that influences the quality indexes of the bridge. If the system of foundations is not defined by the geological conditions and may be changeable, then in the majority of cases this parameter is one of the most substantial. The second alternative is a changeable type of foundation. It is possible that it will be necessary to change the size of the span if the new type of foundation differs substantially in cost from the type of foundation in the first alterantive. A comparison of the two alternatives follows. It may be possible at once to choose the most expedient alternative, which is investigated in relation to the next parameter. This is usually the size of the span, which substantially affects the cost. After that the size of the span is changed, for example, by increasing it,

12.4 THEORETICAL BASIC METHODS OF PRELIMINARY DESIGN

thus composing the third alternative. After that follows a comparison of alternatives. If the third alternative is better, then it is necessary to compose a fourth one by increasing the span once more; if the third alternative is worse, then in the fourth alternative the span is diminished.

The comparison of the first, third, and fourth alternatives provides the possibility of accepting one of them, for example, the third one. In this one it is necessary to investigate the type and construction of span structure. To do this we developed subalternatives, because the basic alternatives are already determined. During comparison of these subalternatives typical structures are used.

The comparison of subalternatives permits us to choose the final alternatives at the given bridge system. After this another system is developed, for example, the arch type. The comparison of the best alternatives of both systems results in the best project for the bridge from the given material. For another material, the whole investigation is repeated. Finally, for each bridge material the best solutions are obtained, a comparison of which gives the final choice of material and bridge project.

An example of the design process is shown below. This scheme is given only to choose alternatives to a reinforced concrete bridge and shown only to clarify the process of design.

Example of Alternatives for a Preliminary Project of a Highway Bridge (Tables 12.1 and 12.2). Basic data: clearances, design loads, location, span of the bridge, and type of foundation (wells) are given. Material: reinforced concrete.

12.4.3 Choice of Final Alternative of Reinforced Concrete Bridge

For given material (reinforced concrete), the third alternative is chosen. In this method of reasoning, from the point of view of theory there is the following

TABLE 12.1

	Beam System
First alternative	System of span structure, deck-type beam. Bridge having three spans. Construction of span structure—from reinforced concrete having four main beams. Supports are massive.
Second alternative	Same, only two spans.
Third alternative	Same, only four spans.
Comparison of alternatives	The best alternative is four spans (third alternative). The first and second alternatives are cancelled.
Subalternatives of third alternative	(a) Four-span beam with two main beams and supports from two columns. (b) Same, with prestressed concrete. (c) With application of welded reinforcing frame.
Comparison of alternatives	The third alternative is chosen: four-span bridge with two main beams from reinforced concrete.

TABLE 12.2

	Arch System
Fourth alternative	Arch type, three spans with four separate arches and columns above arches. Supports are massive.
Fifth alternative	Same, two spans.
Sixth alternative	Same, four spans.
Comparison of alternatives	Fourth alternative chosen: three-span bridge with four separate arches.
Subalternatives of fourth alternatives	(a) With two narrow arches and walls above arches. (b) With box-type arches.
Comparison of alternatives	Fourth alternative chosen; three-span bridge with four separate arches.

defect: In the mathematical problem that is used for the analogy, it is necessary to solve (12.2) together. However, here each one actually was solved separately: First one parameter was found, then leaving it as a constant, another was found. For example, after determination of the most advantageous span size for a certain type of construction, the span structure was varied for its type of structure, trying to find the best. A legitimate question is: For different types of span structure will the most beneficial span size be different? Such a question undermines the security of correctly finding the best solution. We should consider this a theoretical inconsistency. In the case of doubt it is necessary to continue checking by the method of successive approximation. If, for example, changing the type of construction of the span raises doubt about the correct choice of span size, it is necessary to determine it again using the new type of construction.

Generally it is necessary to consider theoretical error, because the initial values of parameters used are not arbitrary, but based on practice, are close to the right values. The order of investigation of separate parameters is also important. At the beginning we investigate the parameters most important for the quality indexes of the structure; next ones of smaller value. For example, type of structure (monolith or prefabricated, conventional or prestressed, or reinforced concrete span structure) has a smaller value for cost than the span size. If the following parameters are used incorrectly the first time, this has a small influence on the preceding. Apart from this, it is necessary to consider that all parameters may change. Therefore, when the next parameter is changed, the preceding most often becomes the former. Bridge span may be changed only by changing the number of spans. For example; if by relieving the weight of the span structure, it is necessary to decrease the span, or it is necessary to change designs to a bridge with a smaller number of spans.

Finally, the most important benefit of applying theoretical methods, is that the whole process of design is not abstract, but is based on real materials. Therefore, in the process of design exists the possibility of introducing separate corrections. Apart from this, after all alternatives are examined together, what is analogous solves (12.2).

A proposed design scheme is characterized by finding the best solution through detailed investigation for each material and for each bridge system; after this, the best solutions are compared. The number of alternatives obtained is large. In the scheme of variation given in Table 12.3 it was necessary for each bridge system to compose three alternatives with subalternatives, for a total of 12. Such difficult investigation is not always necessary. In some cases shorter methods may be applied. In the example in Table 12.3 it is possible to choose bridge material first, thus composing one alternative of steel and one of reinforced concrete.

It is necessary to consider information from practice that uses the most conventional bridge schemes. From their comparison the material is chosen, and from the alternative chosen, by the material, alternatives of that material choice are composed. The bridge system is determined by comparing it with the previous alternative. The accepted alternative is investigated considering the span and other parameters as in the previous cases (see the scheme given below).

Preliminary Design of Highway Bridge. Initial data: clearances, design loads, location, bridge span, and type of foundation (wells) are given. Materials: reinforced concrete and steel.

The difference between the shortened method and the previous one is in the greater use of practical data and in the smaller number of investigations, and

TABLE 12.3

First alternative	Reinforced concrete beam, three span bridge of deck system, with four main beams. Supports are massive.
Second alternative	Steel beams, two spans.
Comparison of alternatives	The first alternative is chosen, reinforced concrete bridge.
Third alternative	Reinforced concrete arch, two spans having four separate arches.
Comparison of alternatives	After comparison of the first and third alternatives, the first alternative is chosen—beam bridge.
Fourth alternative	Two spans, reinforced concrete beam bridge.
Fifth alternative	Four spans, reinforced concrete beam bridge.
Comparison of alternatives	After comparison of the first, fourth, and fifth-alternatives, fifth alternative is chosen: four-span bridge.
Subalternatives of fifth alternative	(a) With two main beams, monolith. (b) Same, with four beams, prestressed concrete, prefabricated. (c) Same, with welded reinforcing frame.
Comparison of alternatives	By comparison of the fifth basic and additional alternatives, fifth subalternative b is chosen.

due to a substantially reduced number of alternative projects. It is obvious that the shortened method is applied in the design of many, if not most, conventional bridges for which existing practice data is available resulting in new investigations being unnecessary.

12.4.4 Conclusions

In the given method of design the most important factors are given basic statements and tendencies. Details of the process are given generally for clarification of these basic statements.

The first basic statement consists of the following: The method of design should be scientific, using a given analogy with the problem of variation analysis. From it the designer uses not the formally mathematical, but a logical nature required in all cases of problem solving. Therefore it plays a leading role in the investigation of the design.

The second basic statement is that the design should be closely connected to the practice and that experience should be applied in the design and building of bridges. This statement is expressed by using examples of bridge projects, which were justified in practice for starting objects of investigation. During the process of design and investigation these examples are analyzed and changes are introduced into them, with the aim of improving them. Here is shown the creativity of the designer, expressed by rationalization and invention. These creative searchings are not separated from practice and are an integral part of the preliminary investigations.

For further examination of the basics of proposed methods it is possible to compare it with the method, developed in detail as creative method. As is indicated above, by this method the whole process of design is divided into three stages: composition of scheme, development of preliminary designs, and detailed design. In the first stage a number of schemes is developed.

With the help of such schemes the designer tries to find the best solution from all that appear possible. Sometimes ten or more schemes are necessary. As a result of this work the designer chooses a few alternatives. During their choice he is governed by common sense, existing examples, and artistic sense. During the composition of solutions, such alternatives develop unlimited possibilities for development and application of creative capabilities and inventions.

In the second stage each chosen scheme is developed in preliminary design. Preliminary design is developed of two or three alternative schemes. During development of the preliminary design, investigative work is important. Considering the analogy used, the first stage—composition of scheme—is used to find the most convenient type of the function U. From the very beginning the designer is using a number of such functions and chooses from them the most applicable, on the basis of general considerations. Further, during preliminary design by investigation of separate parameters the best quality index of the bridge is found. Therefore, the first characteristic of the method of creative direction is to find a function U from investigation of its parameters.

The method of using a number of schemes has as its purpose to cover completely all possible solutions. It may be called concentrical. After preliminary separation of schemes the choice is narrowed down and remaining solutions are developed in more detail. This method serves somehow as the basic idea of competitions. Competition mobilizes the creative fantasy of a number of designers, without the hope that a single designer, even with high qualifications, can solve the total problem. However, it is necessary to note some deficiencies of this method for convenient designing.

Bridge schemes at the first stage are only outlined or sketched and are not investigated in detail; separation is carried on the basis of very general considerations and the personal taste of the designer. Methodical investigation is not applied here. Everything is based on the personal creation, schemes of the bridges are not connected to practice, and there is no guarantee that the designer will "outline" and choose the most expedient schemes.

In contrast, in the method proposed in this text, the choice of bridge scheme is connected to the preliminary development, or should be investigated. The most practical scheme is chosen at the beginning and creation is shown later in the change of scheme on the basis of investigation. This method may be called "progressive," from one investigated scheme to another. Here logical thinking forces the fantasy of the designer toward improvement of the structure and simultaneously introduces the basis of experimental knowledge and common sense. Systematic and methodical investigation of all parameters guarantees that all problems of design of the structure will be completely covered. Therefore, the basic difference between the method proposed here and the method of creative direction consists in a different approach toward the choice of scheme. There is no difference in the preliminary design of the chosen scheme or the investigation of different parameters.

The difference between the methods of design considered and those of the creative technique is of critical importance and may be applied also to bridge design courses, and also to other problems connected with design.

REFERENCES

1. Waddell, J. A. L., *Bridge Engineering*, vol. I, Wiley, New York, 1916, pp. 267–280.
2. Mitropolskii, N. M., *Methodology of Bridges Design, Scientific-Technical Edition*, Avtotransportnoi Literatury, Moscow 1958, pp. 215–242 (in Russian).
3. Polivanov, N. I., *Design and Calculation of the Reinforced Concrete and Metal Highway Bridges*, Transport, Moscow 1970, pp. 5–36 (in Russian).
4. Steinman, D. B. and Watson, S. R., *Bridges and Their Builders*, Dover Publications, New York, 1957, pp. 378–391.
5. DeMarè, E., *Your Book of Bridges*, Faber and Faber, London, 1963, pp. 11–28.

CHAPTER 13

METHODOLOGY OF PRELIMINARY DESIGN

13.1 INTRODUCTION

The planning and design of bridges are part of engineering art, because each bridge, irrespective of its utilitarian function and detailed analysis, is the manifestation of the creative capability of the designer, as is any work of art.[1,2] In bridge engineering the basic act of creative capability is imagination. To plan and design a bridge it is necessary first to imagine it. However, to imagine a bridge, the designer should possess experience based on previous work and apply his knowledge to local conditions. But the beginning designer may imagine a future bridge on the basis of study and critical considerations of different bridge schemes. Generally the designer approaches his problem successfully in two stages. The first and most important stage consists of the creation of the bridge scheme. Further, this scheme is checked and put on the drawing, because only by the drawing is it possible to imprint and check achievements of imagination.

This checking is carried out regarding application to the local conditions (considering span, construction height, profile, etc.), economy (choice of span structure and configuration), cost and aesthetics (general view and harmonic matching with locality). Because now it is important to obtain results quickly and the only problem is the possible application of the bridge scheme, such as it is quite appropriate here to use methods to simplify the use of drawings, formulas, analogy, and so on. A summary of the two stages above result in the bridge scheme and the real proposal.[3] The scope of these two stages is to prove the possibility and rationality of application of the scheme conceived by the designer. However, because a few rational schemes may exist, they may be compared. First a rough comparison of a few schemes can be made. But because the schemes for solving the problem are generally similar, such a comparison often does not provide a clear solution. Then it is necessary to go to the next stage, namely, the preliminary design, the aim of which is the comparison of ideas.[4,5] And because here similar alternatives are compared, it is necessary to consider the elements of the bridge, such as the deck, span structure, supports, and so on, comparing each part separately and summing the results. First it is necessary to imagine each part, further to draw it, and then to check its rationality, application, and economy, which is performed by the analysis. Therefore the drawing is followed by the analysis, which helps to make corrections.

Because the preliminary design does not provide a ready solution, but only a final choice of alternatives, here the relevant shortening of the work is by checking use of auxiliary coefficients, and so on. However, the differences that the preliminary project should contain usually are not substantial; therefore the design calculation should be exact and complete and the use of some coefficients to change the results should be exactly and definitely applied. Design calculation is done on the basis of structural mechanics. Usually the analysis starts with the deck, stringers, and transverse beams, which determine deck weight. Final analysis includes main carrying bridge members, determination of the forces, total weight, and analysis of the bearings. Parallel with the analysis, correction of the initial construction scheme is carried out. However, at the preliminary design the purpose is only to explain the characteristics of the alternatives. The basic comparison of alternatives is a comparison of the weights of structure and cost. In the preliminary design the weight cannot be estimated correctly and it is estimated on the basis of experimental coefficients.

Finally the chosen scheme should undergo detailed design. The aim of the detailed design is the structure of the bridge. This idea is basic. Therefore, the analysis is subjugated to the structure, which results from the purpose and work of the bridge as a whole. Each detail is imagined first (put on the drawing), analyzed and checked regarding its rationality, and corrected. Also, for each detail the most beneficial alternative should be chosen.

The sequence of analysis of detailed design remains the same as for preliminary design, only it is more complete because it has formal meaning. The design covers the bridge in all its details and each of its part is drawn on the paper. Finally the weight is estimated considering the actual volume of the bridge elements and plotted on the special list called "specifications" or the list of weights. This list is generally the end of the project. Actually this order is not completely correct, because the project may reach a complete finish only during construction. For this reason the designer, from the beginning of the project, should consider the construction problems and in certain important cases should give complete instructions regarding methods of construction.

13.2 GENERAL CONSIDERATION FOR DESIGN OF STRUCTURAL BRIDGE SCHEME

The design of the structural scheme of the bridge, which includes the determination of general dimensions of the structure, number and sizes of the spans, choice of the rational type of substructure and foundations, and materials of superstructure and substructure, is a complex engineering problem.[6] The technical and economical parameters of the structure and mainly the cost of bridge building, its safety, exploitation, convenience, and external view depend upon the correct solution of these problems.

During the design of the bridge, crossing the river should be considered the working cross-section under the bridge, which should provide for the required discharge of the water. The opening of the bridge, which is measured from the level of high water, is obtained at cross-sections between piers considering the configuration of the river channel, coefficient of stream compression, and permissible erosion of the river bed. By changing the erosion coefficient and the cross-sectional area in the limits permitted by the standards, it is possible to obtain the different acceptable dimensions of opening for the same bridge crossing. During the choice of the most expedient alternative it is necessary to consider that diminishing the bridge opening is connected with increased cost of foundation due to the large depth of erosion and the necessity of applying more complicated and expensive structures for stream flow. During the design of such structures as viaducts and overpasses, their total length is usually given. It may be determined by general planning or landscape of the location and relation of the cost of an embankment of great height and the bridge structure.

The design of the bridge usually starts with the development of a series of possible alternatives. By comparing alternatives, considering technical and economic parameters, we try to find out the most expedient solution at the given local conditions. At the present time the development and comparison of alternatives is the only way to find the most expedient solution. The factors influencing the choice of bridge scheme are various and their number is so great that to obtain the direct answer to what bridge scheme is most rational at the given local conditions usually is possible. It is necessary to develop a few alternatives, schemes which are used on the basis of such logical considerations as local conditions (geological, hydrological, shipping, construction, etc.) and applying the creative initiative to the designer to the choice of one or another structural solution. Therefore, providing structural schemes of bridge alternatives is basically a creative action and sometimes also contains the elements of invention. Computers may be used to determine the most advantageous span to find the rational number of girders on the bridge having a top deck or number of panels in the truss, to choose the substructure, and so on. However, to apply computers to the general choice of a rational alternative, considering a comparison of all technical and economical parameters, is impossible. Finding an optimum alternative using different points of view often leads to different conclusions. For example, the alternative may be the most advantageous by cost, but may require great expenditure on metal or requiring special erection equipment, which cannot be obtained. Some alternatives may not satisfy an architectural requirement, considering city bridges. Therefore, in the case of computer use it is impossible still to refute the conventional design method, considering all problems of local conditions, which are practically impossible to put in the program of the computer.

13.3 SEQUENCE OF THE WORK DURING DESIGN OF BRIDGE ALTERNATIVES

Finding the most expedient solution during development and comparison of alternatives is done as follows:

1. At the start it is necessary to do the general estimate of local conditions and requirements—geological, hydrological, shipping, and other. The estimate and analysis of local conditions are necessary to provide general solutions to the basic equations of construction: choice of type of foundation at shipping and nonshipping spans; construction of foundations, choice of span structures for the main opening and flood lands of the bridge; construction of piers and abutments for connections with the embankments.

2. *Development of Bridge Alternatives and Their Comparison.* After types of superstructure and substructure are established at given local conditions it is necessary to design alternate bridge schemes and indicate approximate general dimensions of superstructure and substructure (spans, height of the beams or arches, widths of supports and others).

3. *Estimates of Alternatives and Their Comparison.* Comparison of alternatives is made by their economical effectiveness considering technical and economical parameters and local conditions. At the present time the alternatives are compared by the following basic technical and economic parameters:

    a. By the cost, considering labor work, time of construction, and also the difference in the cost of exploitation.

    b. By labor work, or general number of man-days necessary for the bridge construction.

    c. Regarding continuity of construction, the volume of concrete, weight of metal for basic and auxiliary structures, simplicity or complexity of the work during bridge erection, external view of the bridge, which may have decisive meaning are considered.

During the preliminary design the alternatives are appraised by the three basic parameters indicated above. The choice of building materials depends on the summary of the technical and economical parameters and the following requirements: economical, construction, production, exploitation, and architectural.

13.4 LOCAL CONDITIONS AND SOLUTION OF GENERAL PROBLEM OF CONSTRUCTION

During the assignment of construction schemes of bridge structure, the following local conditions are considered, which characterize the location of the crossing:

- Longitudinal profile along the axis of the bridge
- Geological data
- Hydrological data
- Opening of the bridge, which is determined on the basis of hydraulic calculations. (During development of the alternatives of the bridge construction it is usually given.)
- Required navigational clearance dimensions under the bridge, if applicable
- Widths of the roadway and sidewalks

1. *Longitudinal Profile of Bridge Crossing.* Longitudinal profile along the bridge axis crossing provides the location of the maximum depth along the width of the river and in the bridges across navigation rivers longitudinal profile permits the necessary disposition of shipping spans along the length of the bridge crossing. Longitudinal profile also provides the elevations of the shores on which depend the longitudinal slope of the roadway on the bridge. By placing shipping spans at the middle of the river and at a symmetrical longitudinal profile of the roadway, the whole structure obtains symmetrical character (Figs. 13.1–13.4) where symmetry satisfies all the conditions of the site.

2. *Geological Data.* The geological section along the bridge axis permits decisions about possible systems of span structures and types of foundations considering statically indeterminate support reactions.

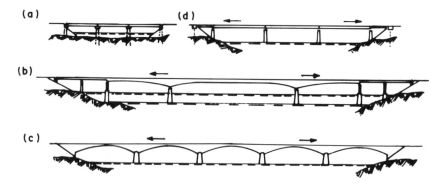

Figure 13.1 Schemes of the reinforced concrete bridges having symmetrically arranged spans.

13.4 CONDITIONS AND SOLUTION OF GENERAL PROBLEM OF CONSTRUCTION

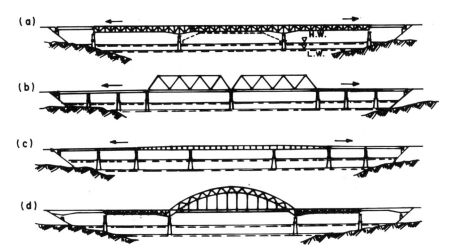

Figure 13.2 Schemes of steel bridges having symmetrically arranged spans. Approach spans are from reinforced concrete.

The simplicity or complication of the foundations to a great degree determine the cost of supports and the whole bridge structure. For this reason the choice of foundation type and its association with the applied system of span structure has a deciding role during compositions of alternatives for the bridge structure. The type of bridge foundation is chosen starting from the geological condition of the crossing location considering methods of the construction requirements. At the present time the tendency is to eliminate deep foundations such as caissons and deep wells, which provide safety but are complicated to build and require a large quantity of material and labor force. A deep pile

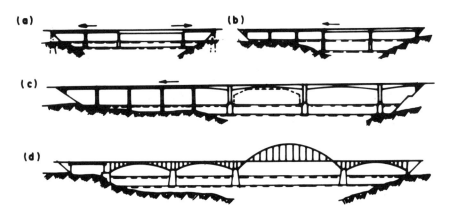

Figure 13.3 Schemes of reinforced concrete bridges with asymmetric arrangement of spans.

230 METHODOLOGY OF PRELIMINARY DESIGN

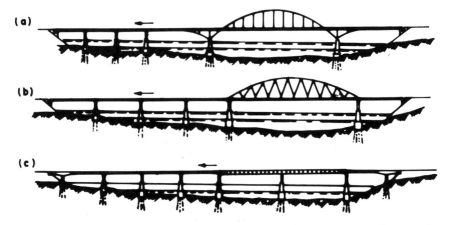

Figure 13.4 Schemes of steel bridges with asymmetric arrangement of spans. Approach spans are from reinforced concrete.

foundation applying reinforced concrete piles and pipe-type reinforced concrete and steel piles and also pile-shells having diameters from 3 ft (1 m) to 20 ft (6 m) are widely used whenever conditions permit.

On the crossings over large rivers geological conditions along the river width are often variable. Due to this, river channel piers have different foundations from those used for the shores on flood lands. For example, piers in the channel are built on the high piles and shore and flood lands on the low piles; or channel piers are built on deep foundations and flood land foundations on piles. The top of the foundation is usually placed lower than the elevation of low water, and at locations that are not covered by water, at the elevation of ground surface after erosion. At piers of shipping spans, minimal water depth for passing of ships close to piers should be insured.

Together with the chosen type of foundation the geological profile also indicates the elevation of the foundation bottom.

3. *Hydrological Data.* The elevation of low water may indicate depths of the river and is not favorable for shipping. Data regarding river depth helps to locate shipping spans along the width of the river. However, in rivers with an easily eroded river channel it is necessary to consider possible shifting with time of most deep parts of the river and together with them the ships' movements along river width. From the elevation of high water are measured the heights for the shipping clearance of the bridge as well as the calculated width of the bridge opening.

It is necessary to know the elevation and intensity of the ice pressure for correct choice of pier structure. For example, at piers of great height parts of the pier built above the elevation of ice pressure can be made of lighter construction as columns. At small ice pressure, and at the small height of the bridge, piles, columns, or a single column may be used for supports.

13.4 CONDITIONS AND SOLUTION OF GENERAL PROBLEM OF CONSTRUCTION

4. *Bridge Length, Opening, and Clear Span.* Bridge length L is the length of superstructure. The opening of the bridge L_0 is the distance between the walls of abutments, measured at the high water level or $L_0 = l_i b_1 + b_2 + \cdots a_1 + a_2 \cdots$. The same distance less the total width of all the bridge piers is called the clear span of the bridge at the level of high water:

$$L_c = L_0 - a_1 - a_2 \cdots = b_1 + b_2 + \cdots = \sum_{i=1}^{i=n} b_i \quad (13.1)$$

where

b_1, b_2 = the individual clear span at the level of high water
a_1, a_2 = width of the piers at the level of high water

13.4.1 Clearance Under Bridges and Underpasses

The clearances given above are defined by the American Association of State Highway and Transportation officials in their *Standard Specifications for Highway Bridges* as follows:

1. *Navigational Clearances.* Permits for the construction of crossings over navigable streams must be obtained from the U.S. Coast Guard and other appropriate agencies.

The channel openings and clearances shall be acceptable to agencies having jurisdiction over such matters. Channel openings and clearances shall conform in width, height, and location to all federal, state, and local requirements.

2. *Highway Clearances for Bridges*

    a. *Vertical Clearance.* Vertical clearance on state highways and interstate systems in rural areas shall be at least 16 ft over the entire roadway width with an allowance for resurfacing. On state highways, and interstate routes through urban areas, a 16-ft clearance shall be provided except in highly developed areas. A 16-ft clearance should be provided in both rural and urban areas where such clearance is not unreasonably costly and where needed for defense requirements. Vertical clearance on all other highways shall be at least 14 ft over the entire roadway width with an allowance for resurfacing.

    b. *Width of Roadway and Sidewalk.* The width of the roadway shall be the clear width measured at right angles to the longitudinal centerline of the bridge between the bottom of the curbs. If brush curbs or curbs are not used, the clear width shall be the minimum width measured between the faces of the bridge railing.

    The horizontal clearance shall be the clear width and the vertical clear height for the passage of vehicular traffic as shown in Figure 13.5. The

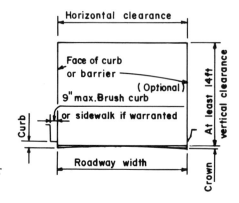

Figure 13.5 Clearance diagram for bridges.

roadway width shall generally equal the width of the approach roadway section including shoulders. Where the curbed roadway section approaches a structure, the same section shall be carried across the structure.

The width of the sidewalk shall be the clear width, measured at right angles to the longitudinal centerline of the bridge, from the extreme inside portion of the handrail to the bottom of the curb or guardtimber. If there is a truss, girder, or parapet wall adjacent to the roadway curb, the width shall be measured to the walk side of these members.

3. *Highway Clearances for Underpasses*

a. *Width.* The pier columns or walls for grade separation structures shall generally be located a minimum of 30 ft from the edges of the through traffic lanes (Fig. 13.6). Clearance diagrams for underpasses where the practical limits of structure costs, type of structure, volume and design speed of through traffic, span arrangement, skew, and terrain make the 3-ft offset impractical, the pier or wall may be placed closer than 30 ft and protected by the use of guardrail or other barrier devices. The guardrail or other device shall be independently supported with the roadway face at least 2 ft from the face of the pier or abutment. The face of the guardrail or other device shall be at least 2 feet outside the normal shoulder line.

13.5 SYSTEMS OF REINFORCED CONCRETE BRIDGES

The types of reinforced concrete beam bridges are shown in Figure 13.7. In Table 13.1 dimensions of the spans and relations of the basic dimensions of these spans are given.[7]

1. *Bridges having small spans (25–50 ft) may be assembled as follows:*

- From T-beams with diaphragms (Fig. 13.7*b*), or without diaphragms (Fig. 13.7*c*).

13.5 SYSTEMS OF REINFORCED CONCRETE BRIDGES

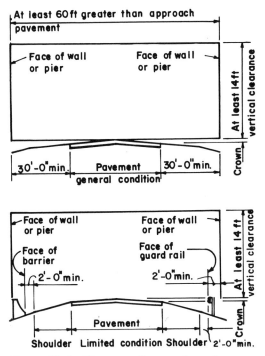

Figure 13.6 Clearance diagram for underpasses.

- From prestressed hollow-core slabs having wire reinforcing (Fig. 13.7*d*).
- From prestressed sections (Fig. 13.7*g*).

The choice among these bridge types depends on a combination of local conditions: Span structures are very simple to build; the main advantage of prestressed core slabs is simplicity prestressed T sections have the lowest erection weight, however, these sections usually require a transverse reinforced cast-in-place structural slab.

2. Most often in short and intermediate span bridges having spans up to 130 ft precast prestressed simple spans are employed. In practice there are cases with application of spans in the range of 165–230 ft. In Figure 13.7 different possible cross-sections of assembled simple span structures are shown: older types with diaphragms between beams (f), newer types without diaphragms (g), and transverse cross-section of spans (h) with more widely placed beams without diaphragms. In the large spans wide placement of main beams is more expedient, which results in a smaller amount of concrete.

3. *Frame-Beam Bridges.* For larger spans, more than 130 ft, frame-beam type bridges are often applied. Their schemes are shown in Figure 13.8*a–c*. Table 13.2 lists general dimensions of frame-beam bridges.

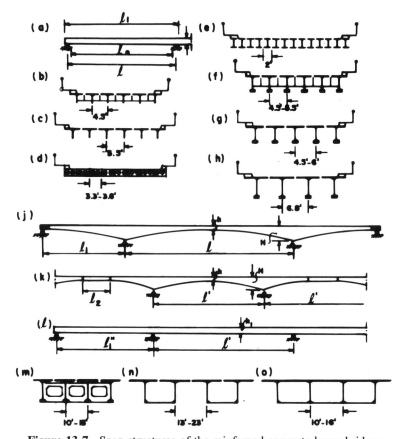

Figure 13.7 Span structures of the reinforced concrete beam bridges.

The main advantage of frame-beam bridges is the possibility of erecting spans without scaffolding by cantilever erection. Also they require less material by comparison with other bridge systems.

The main deficiency of frame-beam bridges is the work of their supports, which apart from compression, work under the moment of fixity of cantilevers, which require a greater quantity of reinforcing, often prestressed.

At the present time the problem of the expediency of application for frame-beam bridges at large spans (295–328 ft) with truss-type reinforced concrete cantilevers is being investigated. Figure 13.8h, i show the basic relations of their dimensions.

4. *Cantilever Spans.* In the number of large bridges having spans of 196–328–490 ft, cantilevered prestressed spans were used (Fig. 13.7i, k). The advantage of such spans by comparison with frame-beam bridges is that their supports are working under vertical loading only in compression, allowing them to be built from reinforced concrete.

TABLE 13.1 Span structures of the reinforced concrete beam bridges

System of Span Structures	Scheme of Construction	Type of Cross-Section	Basic Dimension
1. Simple-span prefabricated reinforced beams	Fig. 13.7a	Fig. 13.7b, c	$l_0 = 25\text{-}50$ ft (7.5-15.0-m) $h \approx \left(\dfrac{1}{11} - \dfrac{1}{15}\right)l_0$
2. Simple-span prefabricated beams having wires as reinforcing	Fig. 13.7a	Fig. 13.7d, g	$l_0 = 25\text{-}50$ ft (7.5-15.0-m) $h \approx \left[\dfrac{1}{19} - \dfrac{1}{23}\right]l_0$
3. Simple-span prefabricated prestressed concrete beams	Fig. 13.7a	Fig. 13.7f, h, m,	10 > 50; 66; 98; 130 ft >15; 20; 30; 40-m $h \approx \left[\dfrac{1}{15} - \dfrac{1}{20}\right]l_0$ For spans $l_1 = 40;\ 50;\ 60;\ 80;\ 108;\ 138$ ft $h \approx \left[\dfrac{1}{13} - \dfrac{1}{20}\right]l_0$
4. Same as under 3, large spans	Fig. 13.7a	Fig. 13.7m	$l = 164 - 230$ ft $l = 50\text{-}70$ m $h \approx \left[\dfrac{1}{18} - \dfrac{1}{20}\right]l$
5. Cantilevered and continuous, prefabricated prestressed concrete beams, erected by the cantilevering method	Fig. 13.7i, k, l	Fig. 13.7n, o	$l = 197\text{-}492$ ft $l_1 = (0.5\text{-}0.7)l$ $l_1^I = 197\text{-}262$ ft $l_2 \approx (0.2 - 0.35)l$ $l_1^I \approx 164;\ 262;\ 295$ ft $l_1^I \approx (0.6\text{-}0.7)l^I$ $h \approx \left[\dfrac{1}{40} - \dfrac{1}{70}\right]l$ or l^I $H \approx \left[\dfrac{1}{12} - \dfrac{1}{17}\right]l$ or l^I $h_1 \approx \left[\dfrac{1}{20} - \dfrac{1}{25}\right]l^{II}$

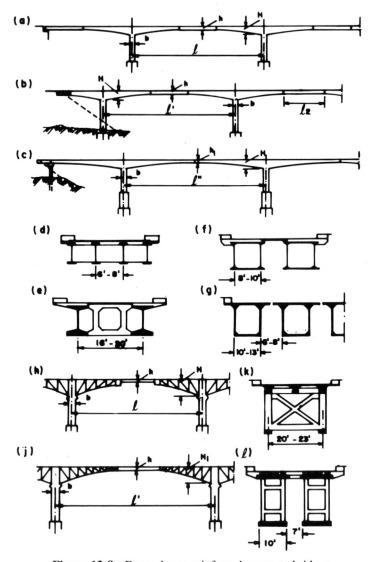

Figure 13.8 Frame-beam reinforced concrete bridges.

5. *Arch-Type Bridges.* At convenient geological conditions reinforced concrete arch bridges may be built, producing thrust (Fig. 13.9a, b, g). At the present time arch bridges are built as assembled structures, having two or three hinges. Data on general dimensions of arch bridges are shown in Table 13.3.

6. *Arches with Rigid Tie.* At spans in the range of 197–262–328 ft it is convenient to build an arch having a rigid prestressed tie (Fig. 13.9g, h). Basic dimensions of the typical assembled span structure of this type are shown in Table 13.3. Structure is combined from separate blocks having a length of no

TABLE 13.2 Frame-beam reinforced concrete bridges

System of Frame-Beam Reinforced Concrete Bridges	Scheme of Structure	Type of Cross Section	Basic Dimension
1. Statically determined and statically indetermined prestressed bridges, assembled or erected by cantilever method	Fig. 13.8a–c	Fig. 13.8d, g, f, m	$l = 130$–460 ft $l^2 = 13$–260 ft $l^{11} = 197$–490 ft $l_2 = [0.3 - 0.4]l^1$ $h \approx \left[\dfrac{1}{15} - \dfrac{1}{50}\right] l$ or l^1 $h \approx \left[\dfrac{1}{40} - \dfrac{1}{60}\right] l^{11}$ $H \approx \left[\dfrac{1}{15} - \dfrac{1}{20}\right] l, l^1,$ or l^{11}
2. Bridges with trussed cantilevers	Fig. 13.8h, i	Fig. 13.7n, o	$l = 295$–328 ft $l^1 = 262$–394 ft $h \approx \dfrac{1}{35} l$ or l^1 $h = \dfrac{1}{9} l$ $H_1 \approx \left[\dfrac{1}{10} - \dfrac{1}{12}\right] l^1$ $b \approx \dfrac{1}{25} l$ or l^1

more than 10 ft. Structure may be assembled on the shore or on the flood part and further moved to the required location. Erection also is possible by barges or on scaffolding. An important advantage of the arch bridge with a rigid tie is small construction height.

7. *Cantilever-Arch Bridges.* Such bridges (Fig. 13.9i, k) have a longitudinal movable hinge in the middle of the span. Thrust is taken by the upper tie, which is formed by the elements of the deck, compressed under prestressing. Bridges of this system require smaller amounts of concrete and high-strength reinforcing than the beam-cantilever bridges, similar by scheme (Fig. 13.7i), but more complicated by erection.

13.6 STEEL AND COMPOSITE BRIDGES

Bridges having spans of steel beams with a composite reinforced concrete deck slab, are shown in Fig. 13.10. The relations of the general dimensions of such spans and also the characteristics of the calculation of steel weight (*a* and *b*) by the formulas of Streletzkii[8] are given in Table 13.4.

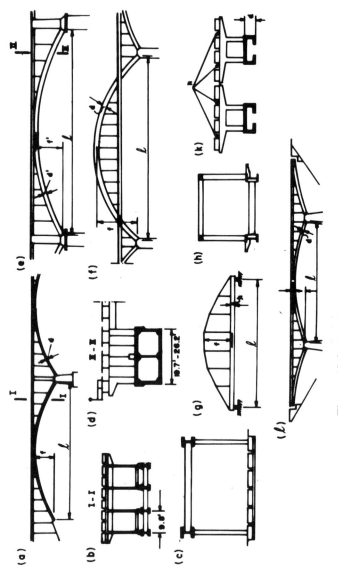

Figure 13.9 Reinforced concrete arch bridges.

13.6 STEEL AND COMPOSITE BRIDGES

TABLE 13.3 Reinforced concrete and bridges

System of Assembled Reinforced Concrete Arch Bridges	Scheme of Construction	Type of Cross Section	Basic Dimensions
1. Three-hinged separate arches	Fig. 13.9a, b	Fig. 13.9c, d	$l = 130\text{–}260$ ft $l^1 = 328\text{–}492$ ft $f = \left[\dfrac{1}{5.5} - \dfrac{1}{7}\right]l$ $f^1 = \left[\dfrac{1}{6} - \dfrac{1}{8}\right]l^1$ $d = \left[\dfrac{1}{4.5} - \dfrac{1}{5.5}\right]l$ $d^1 = \dfrac{1}{50}l^1$
2. Bridge with the traffic in the middle	Fig. 13.9g	Fig. 13.9e	$l = 260\text{–}492$ ft $f = \left[\dfrac{1}{4} - \dfrac{1}{5}\right]l$ $d \approx \dfrac{1}{60}l$
3. Arch with rigid tie	Fig. 13.9g	Fig. 13.9h	$l = 197;\ 262;\ 328$ ft $f = \dfrac{1}{5}l$ $h = \dfrac{1}{35}l$
4. Arch-centilever bridge deck type	Fig. 13.9i	Fig. 13.9k	$l = 197\text{–}459$ ft $f \approx \dfrac{1}{10}l$ $d = \dfrac{1}{50}l$

1. *Steel Plate Girders.* Figure 13.10a–d shows schemes of having plate girders as statical structures (simple, continuous two and three spans, and multiple continuous span). For all of these structures it is assumed that under positive moments they are acting as a composite.[9]

The design spans of typical structures are: simple, 140–210 ft; continuous, 140; 210; 270 ft. At the present time plate girder bridges are widely applied because of the following advantages: economy that is possible with composite design, and simplicity in the fabrication and erection due to simplicity of welding work at the T-sections of steel elements. Structures most often are made from high-strength steel and on bridges of larger spans (more than 390 ft) prestressing and other methods of regulation of forces and stresses are applied. Of the different cross sections alternatives of the plate-girder bridges, the type

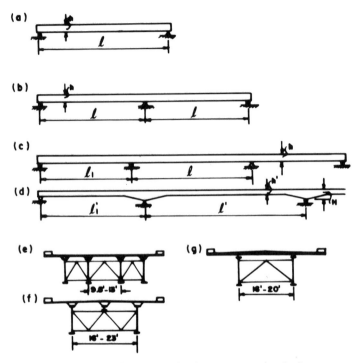

Figure 13.10 Bridges having a composite deck.

shown in Figure 13.10e is applied most often. In structures of high-strength steels the cross-sections shown in Figure 13.10f, g are applied because, owing to the reduction of dimensions of the plate girders, it is possible to obtain economy in the steel.

2. *Deck-Type Trusses.* Figure 13.11 shows schemes of spans with steel trusses, and ratios of their general dimensions and weight characteristics are given in Table 13.5. Deck-type trusses (Fig. 13.11a, b) are applied if construction height permits a substantial increase in the height of the structure. Trusses may be steel (Fig. 13.11g) or composite structures (Fig. 13.11d). A deck-type truss provides economy in steel by comparison with plate girders; however, it requires greater construction height and fabrication.

3. *Through Trusses.* Such trusses may be used for large spans having a small deck construction height (Fig. 13.11f–o). At the present time through trusses are applied for two types of simple spans: (1) loads are transferred at joints and the deck has stringers and floor beams (Fig. 13.11f–l); (2) spans with rigid bottom chords, in which the deck is supported only by the floor beams, which are connected to the bottom chords not only at the joints, but along the panel length (Figs. 13.11m–o). These beams are working as a composite with the slab of the deck. Typical design spans with rigid bottom chord are 207, 273, and 340 ft. At the large width of the channel of the river and

TABLE 13.4 Bridges having a composite deck

System of Span Structures	Scheme of Structure	Type of Cross Section	Basic Dimensions	Weight Characteristics
1. Simple beams	Fig. 13.10a		$l = 130\text{-}200$ ft $h = \left[\dfrac{1}{15} - \dfrac{1}{20}\right]l$	$a \approx b \approx 4.5$
2. Continuous beams Two spans	Fig. 13.10b	Fig. 13.10e–g	$l = 160\text{-}230$ ft $h = \left[\dfrac{1}{20} - \dfrac{1}{25}\right]l$	
Three and multiple spans	Fig. 13.10c, d		$l = 130\text{-}260$ ft $l_1 \approx (0.3\text{-}1.0)l$ $h = \left[\dfrac{1}{20} - \dfrac{1}{25}\right]l$ $l' = 330\text{-}525$ ft $l_{1'} \approx (0.6\text{-}0.8)l'$ $h \approx \left[\dfrac{1}{35} - \dfrac{1}{50}\right]l'$ $H \approx (0.5\text{-}0.3)h'$	$a \approx 4.8$ $b \approx 3.7$

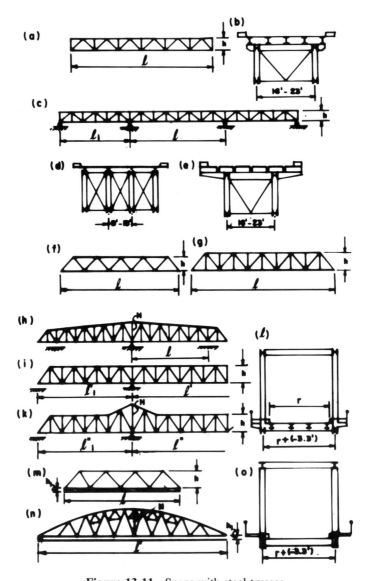

Figure 13.11 Spans with steel trusses.

limited construction height multispan-through bridges are applied, of continuous or cantilever systems (Fig. 13.11*i–k*).

Due to the great rigidity of trusses, application of continuous system is possible if the supports do not settle. The advantage of continuous spans is their ability to be erected without scaffolding or the use of temporary supports. They may be erected by longitudinal pushing, cantilever erection, and other methods.

TABLE 13.5 Spans with steel trusses

Systems of Span Structures	Scheme of Structure	Types of Cross Section	Basic Dimensions	Weight Characteristics
1. Simple span deck system	Fig. 13.11a	Fig. 13.11b, d, e	$l = 100–200\ (230)$ ft $h \approx \left[\dfrac{1}{8} - \dfrac{1}{12}\right] l$	$a = b \approx 4.0$
2. Continuous deck system	Fig. 13.11c		$l = 200–260$ ft $l_1 = 0.71$ $h \approx \left[\dfrac{1}{10} - \dfrac{1}{14}\right] l$	$a \approx 4.2$ $b \approx 3.3$
3. Simple span deck system, flexible bottom chord	Fig. 13.11f, g		$l = 160–230$ ft $l' = 200–330$ ft $h \approx \left[\dfrac{1}{6} - \dfrac{1}{10}\right] l$ or l'	$a = b = 3.3$
4. Continuous or cantilever, deck at the bottom, flexible bottom chord	Fig. 13.11m, n	Fig. 13.11l	$l = 200–300$ ft $l' = 160–330\ (490)$ ft $l'' \approx (0.7–0.8)\, l'$ $l' = 330;\ 490;\ 660$ ft $l_1 \approx 0.751''$ $H \approx \left[\dfrac{1}{6} - \dfrac{1}{1}\right] l$ or l' $h \approx \left[\dfrac{1}{8} - \dfrac{1}{10}\right] l'$ or l''	$a \approx 3.7$ $b \approx 2.9$
5. Simple span, deck at the rigid, bottom chords		Fig. 13.11o	$l \leq 260$ ft $h \approx \dfrac{1}{71}$ or l' $l' = 490$ ft $H \approx \left[\dfrac{1}{6} - \dfrac{1}{7}\right] l$ $h_1 \approx 5\text{–}6$ ft	For Fig. 13.11 $a = b \approx 3.1$ For 13.11 $a = b \approx 3.2$

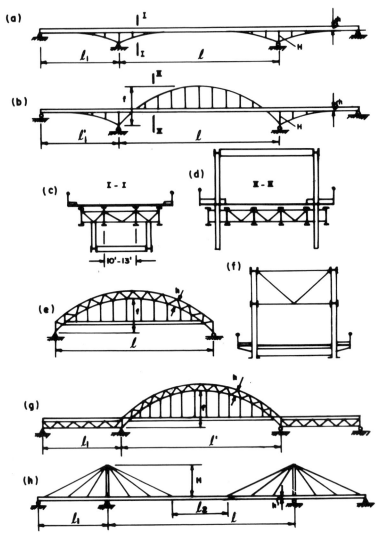

Figure 13.12 Steel spans in combined bridge systems.

4. *Combination of Bridge Systems.* Figure 13.12 shows a combination of steel and reinforced concrete spans, and their characteristics are given in Table 13.6. At continuous girders (Fig. 13.12a, b) it is possible to increase construction height at the supports due to the special configuration of the underbridge clearance. Due to the deloading action of arch, dimensions of the girder and construction height at the middle of span are relatively small or $1/60l$. At erection it is possible first to move the continuous girder, and after that to erect the arch. With a three-spans system, in which the middle arch span is connected to the side girder into the continuous structure (Fig. 13.12f), it is pos-

TABLE 13.6 Steel spans in combined bridge systems

Systems of Span Structures	Scheme of Structure	Types of Cross-Section	Basic Dimensions	Weight Characteristics
1. Beams reinforced by arches	Fig. 13.12a, b	Fig. 13.12c, d	$l = 330\text{–}490$ ft $l^1 \approx 0.55l$ $h \approx \left[\dfrac{1}{50} - \dfrac{1}{60}\right]l$ $H = 5h$ $l_{1'} = (0.4\text{–}0.5)\,l$ $f \approx \left[\dfrac{1}{3} - \dfrac{1}{5}\right]l$	$a \approx 4.6$ $b \approx 3.8$
2. Arch with tie-beam	Fig. 13.12e, f	Fig. 13.12g	$l = 260\text{–}490\ (660)$ ft $f \approx \dfrac{1}{5} l_1\ \text{or}\ l'$ $h \approx \dfrac{1}{20} l_1\ \text{or}\ l'$ $l' = 330\text{–}490$ ft $l_1 \approx 0.5\,l'$	For archs $a \approx 3.8$ $b \approx 2.4$ For beams $a \approx 3.7$ $b \approx 2.9$
3. Cable-stayed system with reinforced concrete stiffening girder	Fig. 13.12h		$l = 330\text{–}525\ (660)$ ft $l_1 \approx (0.4\text{–}0.45)\,l'$ $l_2 \approx (0.25\text{–}0.30)\,l$ $H \approx \left[\dfrac{1}{6} - \dfrac{1}{8}\right]l$ $h \approx \left[\dfrac{1}{60} - \dfrac{1}{80}\right]l$	

sible to build the girder from assembled sections using cantilever erection. In bridges consisting of a continuous girder strengthened by cable stays (Fig. 13.12h), it is possible to build the girder from assembled reinforced concrete sections for spans 460–490 ft, or from prestressed concrete for spans 656–820 ft. Erection of the middle span of cable-stayed bridges may be accomplished by the cantilever method without scaffolding.

13.7 SUPPORTS FOR GIRDER BRIDGES

The basic types of piers and abutments used for girder bridges are shown in Figures 13.13 and 13.14.

1. *Flexible or Column Piers.* For small and medium span bridges and overpasses of relatively small height, when the top of the supports is above the ground level or the bottom of the river is no more than 20–23 ft and in the absence of large ice flow, flexible piles or column supports (Figs. 13.13a, b) are applied most often. The cross-sections of the piles or columns are 14 × 14 in. or 14 × 12 in. At spans up to 33–40 ft the supports are in one row, and at the large spans of 50–66 ft, two rows. Abutments are also in the shape of single-row piles or columns (Fig. 13.14b), which provide the most complete standardized elements of support construction.

2. *Monolith Piers.* Monolith piers are applied mainly in large-span bridges needing great support. Monolith piers are assembled either from solid or hollow concrete blocks of reinforced concrete (Fig. 13.13d). Solid piers usually have a streamlined shape with round or sharp-edged side surfaces. Assembled monolithic piers may have vertical edges and a very high step-type shape (Fig. 13.13d). Also, their side edges may have uniform slope from 30:1 to 50:1 from the top of the foundation to the top of the pier (Fig. 13.13c).

3. *T-Shape Piers.* Modern reinforced concrete and steel girder bridges having spans up to 160–200 ft use piers with cantilevered tops (Fig. 13.13f, g), which provide economy in material by comparison with the solid piers of up to 40–50%. They may be monolith (Fig. 13.13f) or assembled (Fig. 13.13g). The main deficiency of the T-type piers is the substantial amount of steel needed for reinforcing the cantilevers. Apart from this, the blocks of assembled cantilevers have large erection weights, which require powerful erection cranes and a somewhat complicated organization of the work. To reduce the erection weights of the cantilevers different methods for their separation on the blocks were proposed (Fig. 13.13g), such as separation by the longitudinal and transverse joints on four blocks, and separation by wide longitudinal joints on two blocks with filling of the joints by concrete.

4. *Piers with Light Upper Part.* At great heights the intermediate piers are built with a light top part (Fig. 13.13l). The massive bottom part of the pier up to the high-water is given a streamline shape, and the upper above-water part is built from the separate columns of the monolith or assembled construc-

13.7 SUPPORTS FOR GIRDER BRIDGES 247

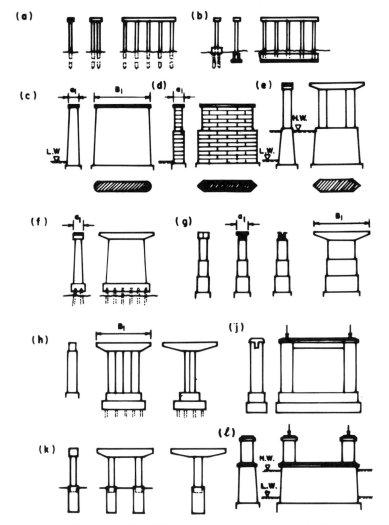

Figure 13.13 Different pier types for girder bridges.

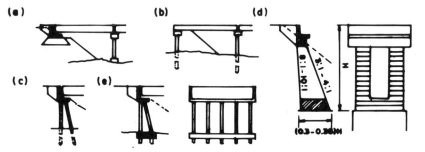

Figure 13.14 Differnt types of abutments.

tion, which support reinforced concrete cantilevers. Disposition of the columns should correspond to the disposition of the girders or trusses in the cross-section of the bridge. However, at spans up to 65–130 ft the minimum number of columns is used: two at the bridges having two traffic lanes.

5. *Column Supports.* In the absence of or for small intensity of ice flow, and at overpasses and viaducts, column supports, having one or two columns, are often applied (Fig. 13.13*h*, *i*, *k*). Columns may be supported by spread footings, pile foundations (Fig. 13.13*h*, *i*) or on hollow piles of large diameter filled with concrete (Fig. 13.13*k*).

6. *Approximate Dimensions of the Piers.* During design of the bridge scheme, the approximate width of the piers (Fig. 13.13*a*) at the top (at two bearings) may be used:

For span 33 ft	3.28 ft
For span 197 ft	6.56 ft
For span 328 ft	9.84 ft

For continuous and cantilever bridges having one bearing at each pier, dimension a_1 is smaller than at a simple span by about 20–25%. The length of the pier on the top transverse to the bridge, $B1$ (Fig. 13.13), is

$$B_1 = B_0 + b\,(1.3\text{–}1.6\text{ ft})\,2 \tag{13.2}$$

where

B_0 = distance between axis of the edge beams
b = width of the pedestal (2.0–4.0 ft)
1.3–1.6 = distance from the edge of pedestal to the edge of support (ft)
Edges of foundation = 1.0–1.3 ft

7. *Different Methods of Junction of the Bridge with the Embankment.* These methods are shown in Figure 13.14. The simplest method of function is to support the end of the span on the lowest abutment, supported by the gravel that is located in the body of the cone Fig. 13.14(a). To eliminate settlement of support layers the ground under the gravel should be carefully prepared. Design stresses of 7 psi are used during transfer of reaction on the filled ground.

Short spans of 33–40 ft may be also supported on pile or column supports located in the body of the cone (Fig. 13.14*b*). A better method of junction is the arrangement of inclined piles (Fig. 13.14*c*) or columns (Fig. 13.14*d*). These supports are used at spans of 100–130 ft, and embankment heights up to 23–26 ft. Abutments in the ground as supports for large spans and reactions are often used (Fig. 13.14*e*). They may be solid concrete retaining walls, placed in the cone body, or may be of column construction. Columns in this case are constructed of reinforced concrete. The dimensions of such abutments are shown in Figure 13.14e. They are applied at the height of the embankment from 6.5–10 ft to a maximum height of 66–82 ft.

13.8 DESIGN OF THE SPANS

Bridge spans are designed considering the following basic requirements:

- Providing under-the-bridge the navigational clearances corresponding to the accepted class of the river
- Maximum economy
- Possible standardization of construction elements of the bridge

Therefore, dimensions of spans to be designed depend on minimal cost of the span structure.

1. *Size of Beneficial Span.* This depends on the type of foundation of bridge supports and applied system of span structure. For girder bridges the most beneficial size of the span approximately corresponds to cases where the cost of the support is equal to the cost of the span less the deck. For an arch bridge a theoretical expression for most beneficial span and its size may be obtained and is usually greater than that for the girder span. However, theoretical formula estimates of size are only approximate. The local conditions may substantially change its dimensions. For a more reasonable determination of beneficial span it is necessary to prepare a few alternatives and to compare them using the cost and other technical and economical parameters, considering local conditions.

2. *Design of Spans for Reinforced Concrete Bridges.* If due to the geological conditions at the supports it is impossible to transfer thrust, a girder system may be used—simple, continuous, and cantilever (Fig. 13.7), frame-beam bridges (Fig. 13.8), or arches with ties (Fig. 13.9). An economically beneficial solution is when all spans are equal (Fig. 13.15).

If the economically beneficial span obtained is smaller than required for shipping or if the bridge length does not permit equal spans, then it is necessary to use more complicated schemes with unequal spans (Fig. 13.16).

Because of the great difference in the dimensions of shipping and flood spans it is not possible to use girder construction (Fig. 13.16a, b), utilizing instead a shipping span arch system that takes thrust (Fig. 13.16c). When laying out spans for girder bridges of assembled construction (simple, cantilever, or continuous), the proposed methods of erection and existing erection equipment should be considered. The dimensions of the spans should be chosen so that the weight of the erection elements corresponds to the power of the existing means of transportation and erection cranes.

When laying out spans for frame-beam bridges, equal spans are most often

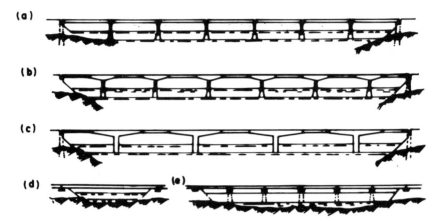

Figure 13.15 Examples of reinforced concrete bridges having unequal spans.

applied (Fig. 13.15c), which provides maximum standardization of elements. However, the application of unequal spans is also possible (Fig. 13.16).

By changing span systems, number of spans, their combination, and dimensions, it is possible to gain a number of alternatives, satisfying given local conditions.

3. *Designs of Spans for Arch Bridges having Thrust.* Arch bridges may be deck arch, deck-through arch, and tied arch. Depending on the local conditions nonhinged, three-hinged arches, or arched bridges of combination systems such as rigid beam with flexible arch may be applied. The last two systems may be applied at assembled bridges with great success. The most rational alternative

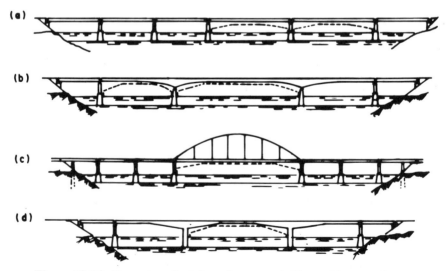

Figure 13.16 Examples of reinforced concrete bridges with unequal spans.

construction of the arch bridge, if permitted by local conditions, is the bridge with equal spans and deck-type arch. In this case the thrusts on the piers are equalized under dead load and standardization of span elements and supports is insured, which is especially important at assembled bridges (Fig. 13.17a). Such bridges may also be used for shipping spans of large dimension and shore spans of smaller dimensions (Fig. 13.17b). In this case equalization of the thrusts under dead load dimensions of their rises should be found as follows:

$$f_1 : f_2 = p_1 l_1^2 : p_2 l_2^2 \tag{13.3}$$

where p_1 and p_2 are dead load intensities per foot. Equation (13.4) follows from the thrust of the three-hinged arch:

$$H = \frac{pl^2}{8} \tag{13.4}$$

For equalization of thrusts at spans (Fig. 13.17) it is possible to apply large spans at lighter constructions, for example, and separated arches and side spans at heavier solid valuts. When there are great differences in the dimensions of shipping and flood spans, it is possible to equalize thrusts applied at the shipping span deck through an arch in the middle (Fig. 13.16c), which permits a substantial rise in the arches and thus reduces thrust.

When ground conditions are poor, it is possible to apply equal spans having deck-through arches along the whole length of the bridge (Fig. 13.17d), to reduce and standardize the thrust transferred to the support under the side live load.

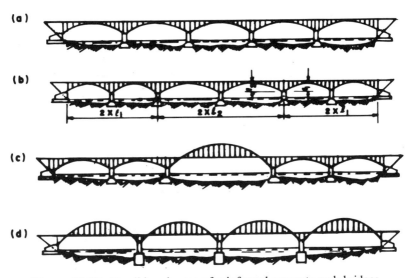

Figure 13.17 Possible schemes of reinforced concrete arch bridges.

During the development of schemes of assembled arch bridges, the possibility of separating them on erection elements should be considered, which is convenient for transportation and erection, the weight of which corresponds to the carrying capacities of transportation means and erection cranes.

4. *Considerations Regarding the Spans of Steel Girder Bridges.* Dimensions of spans similar to reinforced concrete bridges are necessary when considering dimensions of navigational clearances and when the cost of the span together with support must be minimized.

Figure 13.18 shows spans for the case when the required navigational span is 197 ft, but the economically beneficial span is equal to 230 ft. Apart from the economic advantage, Figure 13.18a is also preferred as a standardized span structure. Often when laying out bridge spans there are cases when navigational factors require the use of steel spans and remaining spans are reinforced concrete, which are not economical.

Figure 13.18b shows three shipping spans (in the case of shifting channels), which are spanned by continuous steel trusses, having rigid bottom chords (see Fig. 13.11m). The flood plain of the river is spanned by reinforced concrete, typical assembled structures.

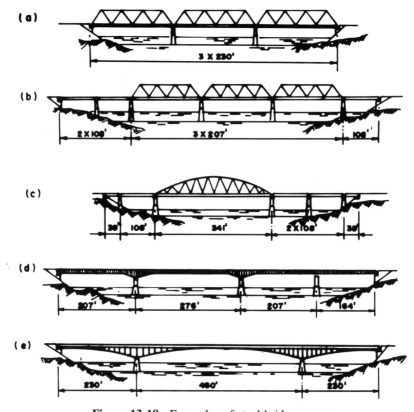

Figure 13.18 Examples of steel bridges spans.

Figure 13.18c shows another case of the bridge having steel and reinforced concrete spans. The navigational opening has a width of 328 ft and is spanned by the segmental steel truss having rigid bottom chords (Fig. 13.11h), and the remainder of the bridge has reinforced concrete spans. Cones are covered by small spans, which are supported by piles placed in the body of the embankment. Different systems of steel spans may be successfully applied under different local conditions. Figure 13.18c also shows alternative bridge constructions with the upper deck over the large navigational span and small construction height.

By changing the number of spans, their combinations and dimensions, and their systems, it is possible to design a series of alternatives, satisfying given local conditions, and by comparing their technical-economical indexes, to find the most expedient solution.

13.9 DETERMINATION OF THE AMOUNT OF BASIC WORK DURING THE DESIGN OF BRIDGE ALTERNATIVES

To compare alternative structures it is necessary, with some degree of approximation, to determine the quality of materials (concrete, reinforced concrete, reinforcing, structural steel, etc.), but not to make a detailed design of all bridge elements. During design of alternatives basic dimensions from the practice, using average values, are usually assumed. During calculations data about material spending at projects having typical spans and supports are used. In some cases preliminary design calculations are made for separate elements. Due to the relative simplicity of reinforced concrete shapes at spans and supports, calculating their volumes is not difficult if their dimensions are given or determined from preliminary calculations. The amount of steel needed at the steel spans is determined from the special half-empirical formulas.

13.10 EXPENDITURE OF MATERIAL IN REINFORCED CONCRETE, PRECAST SIMPLE-SPAN BRIDGES UP TO 130 FT

1. To determine the quantity of concrete in precast reinforced concrete spans of highway bridges it is possible to use the diagrams shown in Figure 13.19. The volume of concrete is given in cubic feet per square foot of the bridge deck: 1 and 2—with reinforcing; 3, 4, and 5—prestressed beams; 6—multibox beams; 7—I-beams; 8—multibox beams, T-beams. The expenditure of reinforcement in pounds per cubic feet for concrete (Fig. 13.19) is:

For spans of type 1 and 2	550–607 lb
For spans of type 3, 4, and 5	324–354 lb
For spans of type 6 and 7	182–243 lb
For spans of type 8	423–263 lb
For spans of type 9	304–324 lb

254 METHODOLOGY OF PRELIMINARY DESIGN

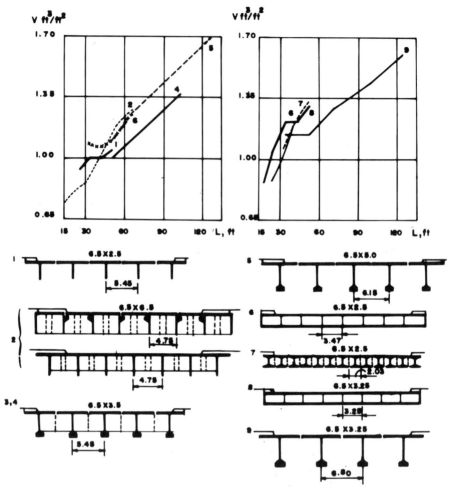

Figure 13.19 Diagrams of expenditure of concrete for precast reinforced concrete spans up to 130 ft.

2. The total amount of concrete and reinforced concrete in different types of massive piers—precast, monolith, precast-monolith—may be determined by the diagram shown in Figure 13.20. Pier No. 1 is precast; No. 2 is monolith; No. 3 is precast monolith.

    I. Pier No. 1 at $l_0 = 50$ and 65 ft
    II. Pier No. 1 at $l_0 = 100$ and 130 ft
    III. Piers No. 2 and 3 at $l_0 = 50$ and 65 ft
    IV. Piers No. 2 and 3 at $l_0 = 100$ and 130 ft

where $a_1 = 3$–3.6 ft; $a_2 = 3.6$–5.0 ft; $a_3 = 4.3$–6.2 ft.

13.10 EXPENDITURE OF MATERIAL IN PRECAST SIMPLE-SPAN BRIDGES

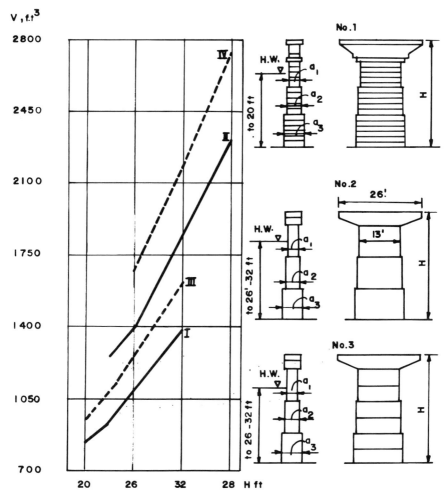

Figure 13.20 Diagram indicating volume of concrete and reinforced concrete used in the function form height of massive piers (H).

The quantity of reinforced concrete in massive piers is as follows:

$$\begin{aligned}
\text{Pier No. 1 at } l_0 &= 65 \text{ ft} & &= 335 \text{ ft}^3 \\
l_0 &= 90 \text{ and } 130 \text{ ft} & &= 388 \text{ ft}^3 \\
\text{Pier No. 2 at } l_0 &= 65 \text{ ft} & &= 318 \text{ ft}^3 \\
l_0 &= 90 \text{ and } 130 \text{ ft} & &= 423 \text{ ft}^3
\end{aligned}$$

The weight of steel for massive piers per cubic foot of the reinforced concrete on average is 83–176 lb.

3. In massive intermediate piers with a superstructure, the quantity of consumed concrete and reinforced concrete may be calculated considering the geometric dimensions of piers as shown in Figure 13.21.

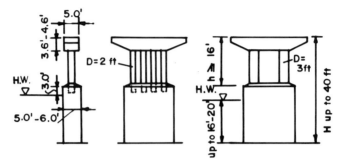

Figure 13.21 Massive piers having a superstructure.

4. A diagram indicating the quantity of consumed and reinforced concrete used for large prestressed concrete simple spans is given in Figure 13.22.

5. The volume of concrete work for the piers for beam spans is the summary of volumes of separate parts (Fig. 13.23) using the formula

$$V = l'B\Sigma h; A; \quad (13.5)$$

where

l' = half summary of spans, from both sides of pier
B = width of the bridge between railings
$h;$ and $A;$ = height of separate parts of the pier and corresponding coefficients A

6. For the volume determination of massive abutments for large-span beam bridges the diagrams shown on Figure 13.24 may be used. The volume of massive abutments having wing walls (I) or only breast walls (II), is

$$V = V_{ab}H_{ab}B \quad (13.6)$$

13.11 WEIGHT OF STEEL IN STEEL AND COMPOSITE SPAN BRIDGES

The weight of a steel span, necessary for the design, is determined from earlier projects using theoretical formulas.[10] The weight of the steel is determined from the weight of the stringers, floor beams and sidewalks, main carrying members (girders or trusses), and bracing.

The weights of steel in stringers and floor beams per 1 ft² plane of the bridge are:

- Floor beams from 8 lb/ft² to 12 lb/ft²
- Stringers and floor beams from 16 lb/ft² to 25 lb/ft²
- Weight of steel for sidewalks from 6 lb/ft² to 10 lb/ft²

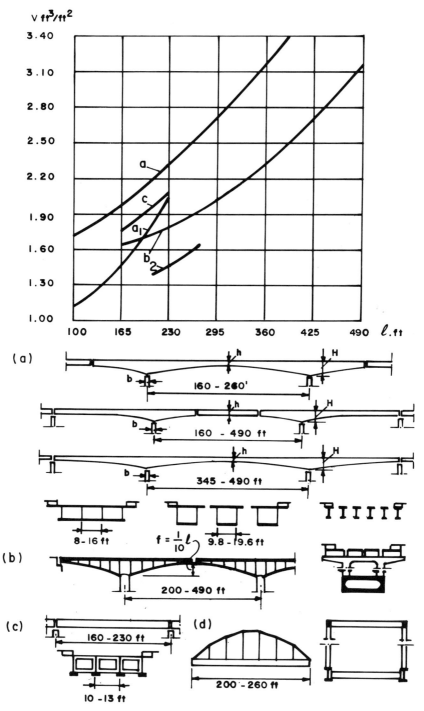

Figure 13.22 Diagram indicating consumed reinforced concrete for large prestressed spans: (*a*) and (*a*1)—cantilever and continuous beams, average values and lowest; (*b*) cantilevered arches with the upper chord tied; (*c*) simple beams of large spans; (*d*) flexible arches with a rigid tie beam.

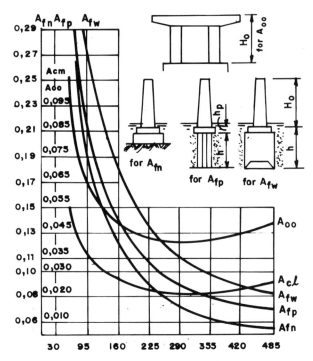

Figure 13.23 Coefficients for volume determination of concrete-work for beam bridges: A_{oo} = for massive parts of the piers above foundation; A_{co} = same for column supports; A_{fn} = for foundation on natural ground; A_{fp} = same for pile foundation; A_{fs} = shaft or caisson.

The weight of steel for bracing is usually a function of the weight of the main carrying members:

$$g_{br} = \delta g_{to}$$

where δ = 0.1–0.12. The most useful theoretical formulas is that proposed by Streletzkii[10], which expresses the weight of trusses or girders as follows:

$$g_{tot} = \frac{an_{LL} + [q_d + (q_{s+fb} + q_{br})]b}{\dfrac{R}{\gamma} - bl} 1 \qquad (13.7)$$

where

n_{LL} = design live load, considering the coefficient of overloading
q_d = design load from the weight of the deck
$q_{s+fb} + q_{br}$ = design load under the weight of the stringers, floor beams, and bracing
l = design span of the truss or girder

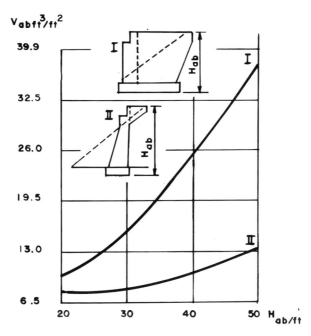

Figure 13.24 Volume of concrete work of massive abutments for beam bridges on natural ground in ft^3/ft^2 considering abutment projection on the vertical plane perpendicular to the bridge axis.

γ = specific weight of the steel
R = allowable stress of the steel
a, b = coefficients that depend on the system and construction of the main carrying system of the bridge (for simple girders, after Streletskii,[8] $a = b = 5$, and for simple trusses $a = b = 3.5$)

The weight of bracings at steel highway bridges constitutes 8–12% of the weight of main carrying members. The dead load of the simple girders is as uniformly distributed along the span. At continuous and cantilever bridges it is necessary to consider the nonuniform weight of main carrying structures along spans.

REFERENCES

1. Vitruvius, *The Ten Books on Architecture*, translated by M. H. Morgan, Dover Publications, New York, 1960, pp. 13–16.
2. Palladio, A., *The Four Books on Architecture*, Stroiizdat, Moscow, 1952 (in Russian).

3. Mitropolskii, N. M., *Methodology of Bridge Design*, Avtotransizdat, Moscow, 1958 (in Russian).
4. Holgate, A., *The Art in Structural Design*, Clarendon Press, Oxford, 1986, pp. 1-6, 24-30, 187-195.
5. Francis, A. J., *Introducing Structures*, Pergamon Press, New York, 1980, pp. 221-260.
6. Waddell, J. A. L., *Bridge Engineering*, vol. I, Wiley, New York, 1916, pp. 267-280.
7. Polivanov, N. I., *Design and Analysis of the Concrete and Metal Highway Bridges*, Transport, Moscow, 1970, pp. 5-58 (in Russian).
8. Polivanov, N. I., op. cit., pp. 57-58.
9. Gibshman, E. E., *Metal Bridges on Highways*, 3rd ed., Avtotransizdat, Moscow, 1954, pp. 315-322 (in Russian).
10. Gibshman, E. E., *Design of Metal Bridges*, Transport, Moscow, 1969, pp. 184-185 (in Russian).

CHAPTER 14

COMPARISON OF ALTERNATIVES

During comparison of bridge alternatives different technical and economic parameters should be considered: cost, labor, length of construction, and amount of concrete and steel for basic and auxiliary parts of the structure.[1,2] When comparing alternative city and main highway bridges, the external view of the structure is important. The choice of alternative may influence the existence or absence of construction means (barges, cranes, pile drivers, etc.). In such cases, the most rational structure, using technical and economic indexes, cannot be accepted and realized if the construction company does not have the transportation facilities and erection equipment required for the work.

14.1 COMPARISON OF ALTERNATIVES BY CALCULATED COST

In the simplest case, when spans, supports, and foundations are the same type and alternatives differ only by dimensions or combination of spans (Figs. 14.1 and 14.2), we consider only calculated costs. It is assumed that the methods and time of construction using each alternative are equal or differ only a little. Approximate data of the calculated cost may be obtained considering amounts of separate types of work.

14.2 COMPARISON OF ALTERNATIVES BY THE COST OF BASIC CONSTRUCTION MATERIALS

When choosing the optimum solution consider also the general cost of basic construction materials (concrete, reinforced concrete, and steel). During the comparison of alternatives by cost of materials, compare:

262 COMPARISON OF ALTERNATIVES

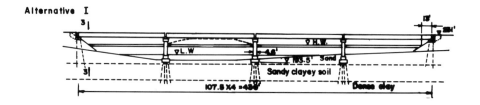

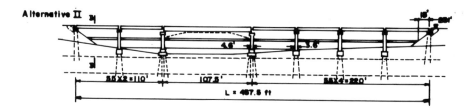

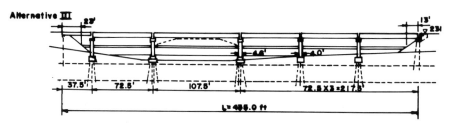

Figure 14.1 Schemes of Alternatives (I, II, III) of reinforced concrete bridges (Example no. 1).

- Volumes of concrete and reinforced concrete
- Weights of different types of steel (high strength wire, steel cable, etc)

If it is possible to calculate the weights of auxiliary members, then the alternatives are also compared as to such auxiliary features of work.

14.3 COMPARISON OF ALTERNATIVES BY CONDITIONS OF FABRICATION AND ERECTION

During the selection of bridge alternatives, to reduce labor and time of construction, preference is given to the structure with the simplest fabrication and erection. Simplification and acceleration of work during bridge construction may be achieved when its elements—reinforced concrete and steel members, as well as supports—are prefabricated, and also when application of the assembled structure, and erection of the system, is done without scaffolding. In

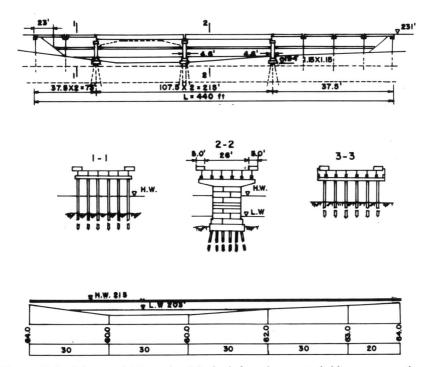

Figure 14.2 Scheme of Alternative IV of reinforced concrete bridge, cross-sections, and longitudinal profile (Example no. 1).

bridges over small rivers, and at overpasses and flood parts of large bridges, it is convenient to apply reinforced concrete spans up to 130 ft erected by cranes, which are moved on the ground. During construction of frame-beam or cantilevered reinforced concrete bridges with spans of 200 ft and more, supports at flood parts of the river are built with the help of portal cranes. Also, in the channel auxiliary scaffolding for pile drivers and cranes are built around the support. Spans are built mostly by the cantilever method. In the arch-type reinforced concrete bridges of assembled construction of spans, supports are monolithic to work against horizontal forces. Arches with 197- to 260-ft spans are often erected from half arches by portal cranes. Arches with spans of 328 ft and more are assembled from short blocks on the single-span steel center. Steel spans across navigation channels are also constructed with minimum possible use of auxiliary construction. During their construction cantilever assembling is applied, erected on the shore with longitudinal pushing on the temporary supports or with floating supports.

Over large depths of the main part of the river channel multispan structures with equal spans (reinforced concrete or steel) may be the most expedient, which are assembled on the shore and transported to the location on floating supports. On multispan bridges, the cost of floating supports is reasonable owing to their multiple uses.

14.4 COMPARISON OF BRIDGE ALTERNATIVES BY CONDITIONS OF PERFORMANCE

For types of bridges except timber, performance expenses are insignificant and cannot influence the choice of alternative structure. During the appraisal of alternatives construction characteristics alter performance, for example, the presence or absence of joints and hinges may influence performance because they are weak places of the bridge structure and require detailed inspection during use and maintenance work. From the point of view of use, continuous bridges in which hinges and joints are absent along spans are preferable to spans with joints. It is necessary to note that the maintenance of properly built reinforced concrete bridges is simpler than that of steel bridges where regular painting must be regularly performed, conditions at riveted joints constantly inspected, and rivets changed periodically. Hence maintenance characteristics often have a decisive value when considering the conditions of performance during the choice of basic bridge alternatives.

14.5 COMPARISON OF ALTERNATIVES BY EXTERNAL VIEW

The external view of bridges—structures of great length located openly, often high above crossing obstructions—may influence the surrounding landscape or architecture of the city. For this reason, during the design of highway and city bridges, great attention is given to the choice of structure type and architectural shapes. However, in modern bridges there is a tendency not to use any decorative "architectural arrangement." Architectural effect is obtained using rational methods of composition of the structure as a whole—through the choice of bridge system and architectural configuration (arch, frame-beam, or girder bridge of variable depth). In city bridges application of through trusses should be avoided because the resulting external view does not easily blend with the architecture of surrounding houses.

Therefore, a comparative appraisal of bridge alternatives from the perspective of external view also has an important influence on the choice of basic alternatives.

14.6 EXAMPLES OF BRIDGE ALTERNATIVES

In the following we consider two numerical examples of a comparison of bridge alternatives. In the first example it is assumed that the bridge has only one navigation opening with a clearance width of 98 ft. For this reason the structure may be composed of different combinations of typical simple spans. In the second example the problem of bridge design is solved by crossing the shipping opening with a steel span. For nonshipping spans, reinforced concrete assembled structures are used when possible.

14.6.1 Example 1. Alternatives for a Reinforced Concrete Bridge with a Clear Shipping Span of 98 ft

The geological characteristics of the river bottom at the location of bridge crossing are important for the proper choice of support foundations, as shown on the scheme of alternative 1 (Fig. 14.1); the longitudinal profile along the bridge axis with hydrological data is given on Figure 14.2. The clear waterway opening of the bridge is 377 ft. The width of roadway is 26 ft and two sidewalks each have a width of 12 ft. The bridge should have one shipping span with a clearance width of 98 ft and a height of 11.5 ft.

Appraisal of Local Conditions and Solution of General Problems of Construction. From the longitudinal profile along the bridge axis (Fig. 14.2) it is clear that the maximum depths of the river are located closer to the left shore and, for this reason, in the majority of alternatives an asymmetric layout of the spans is applied. The geological conditions of the crossing location are convenient because weak ground (sands and clay soil) lay at depths up to 33 ft lower than the level of low water. By driving the reinforced concrete piles 33–40 ft it may be possible to pass weak layers and reach the dense ground and dense clay. As far as the top layers of the ground represent easily eroded sands, it is expedient to place high piles as bases of supports. This provides the possibility of not going deep into the ground, but of building a foundation pit from sheet piling. The elevations of foundations at intermediate piers are the same, because at easily eroded grounds depths may be shifted along the river width.

Assuming that the top of the foundation is placed at 1.60 ft lower than the level of low water and the thickness of the foundation slab at a span of 9.8 ft is 8 ft, we obtain an elevation at the foundation bottom of 19.4 ft. Elevation of the bottom of the span structure at the shipping span is 225 ft. Elevation of the deck at the shipping span, with a girder height of $h = 1/18$, $l = 105/18 = 5.83$ ft and the thickness of the slab is $= 0.39$ ft and elevation of the roadway top is equal to $224.68 + 5.83 + 0.39 = 231.00$ ft. Constructions of intermediate supports for spans of 225.00 and 50 ft are used as assembled from precast concrete with reinforced concrete cantilevers. Using limiting dimensions of cantilevers—about 10 ft—we obtain the transverse beam, the pier width of ≈ 16 ft. Shore supports for some spans are used as piles; those at the heights of the approach embankment of 16–20 ft require a pile length of 30–40 ft. Intermediate and shore supports for spans of 30 ft are used as single-row pile construction.

Alternatives of Bridge Structure Schemes. Different possible alternatives of bridge construction are shown in Figures 14.1 and 14.2. Alternatives are combined from typical assembled simple spans. Widths of concrete piers and distances between their axes are determined from empirical equations (Eq. 13.2). In alternative I all clear spans of the bridge are crossed by four spans at 98 ft; the maximum possible standardization of construction elements and also uniformity in erection methods for the clear span is provided by: $430.34 - 4.6 \times 3 - 2 \times 13.12 = 390.30 > 377.2$ ft.

For Alternatives II and III a span of 98 ft is used only to cross the shipping span. The nonshipping part of the bridge is composed of spans of 50 and 66 ft. In the Alternative III, to obtain the necessary length of clear openings between supports, it was necessary to place a span of 33 ft above one of the cones.

Clear spans are equal:

In Alternative II: $437.22 - 2 \times 4.6 - 4 \times 3.6 - 2 \times 13.12 = 387.38 > 377.2$ ft

In Alternative III: $434.93 - 2 \times 4.6 = 3 \times 3.94 - 22.96 - 13.13 = 377.83 > 377.2$ ft

Alternatives I–III were developed to find the most beneficial span for nonshipping parts of the bridge. Thus it was necessary to compare the cost of materials for these Alternatives. During development of Alternative IV we tried as much as possible to consider conditions of exploitation. Crossing the deep part of the river with two large spans of 98 ft each provided the best way to pass high water and ice. Using two openings of 98 ft also permits a reserve shipping span during displacement of the deep part of the river toward the right shore. Such a change in longitudinal profile is possible if the river bottom is formed by easily eroded grounds. Flood parts of the bridge may be composed from small spans, where $l = 33$ ft, on very economical single-row pile supports. The clear span in Alternative IV is $439.52 - 3 \times 4.6 - 4 \times 1.15 - 2 \times 22.96 = 375.20 \approx 375$ ft.

Amount of Materials

Alternative I

- Volume of reinforced concrete for a span of 107.5 ft is $1.45 \times 107.5 \, (26 + 2 \times 5) \times 4 = 22{,}446$ ft^3
- Volume of precast concrete for intermediate supports with a height above the foundation of about 23 ft is $= 3238$ ft^3
- Volume of reinforced concrete horizontal beams at intermediate supports is 1429 ft^3
- Volume of reinforced concrete of horizontal beams for pile supports at abutments is 1644 ft^3
- Volume of slabs on piles of intermediate supports is 4975 ft^3
- To determine the number of piles of intermediate supports we calculate the vertical load on piles due to the weight of spans with covered decks

$$(1.45 \times 0.142 \times 1.1 + 0.046 \times 1.5) \times 107.5 \times (26.0 + 2 \times 4.82)$$
$$= (0.226 + 0.069) \times 107.5 \times 35.84 = 0.295 \times 3853 = 1140 \text{ kips}$$

- Weight from truck loading and people on the sidewalk:

$$\text{Impact I} = 1 = \frac{50}{107.5 + 125} = 1.215$$

$$\text{Sidewalk P} = \left(30 + \frac{3{,}000}{107.5}\right)\left(\frac{55 - 30}{50}\right) = 29 \text{ lb/ft}^2$$

$$2 \times 0.64 \times 1.215 \times 107.5 + 2 \times 28.0 + 2 \times 5.0 \times 107.5 \times 0.029 =$$
$$167.184 + 56.000 + 31.175 = \text{kips}$$

- Weight of the pier is 484 kips
  The number of piles at the base of intermediate support is

$$\frac{1140 + 255 + 484}{100} \times 1.4 = \frac{1879}{100} \times 1.4 \cong 27 \text{ piles}$$

For three supports, 27 × 3 = 81 piles where 100 kips approximately calculate the carrying capacity of one pile in kips and 1.4 is the approximate coefficient considering performance at the base of the high piles. The loads on piles at the abutment are:

- Under the weight of span

$$0.5 \times 1140 = 570 \text{ kips}$$

- From truck loads and people

$$2 \times 0.64 \times 1.215 \times 0.5 \times 107.5 + 2 \times 28.00 + 2 \times 5.0 \times 107.5$$
$$\times 0.5 \times 0.029 = 83.592 + 56.000 + 15.588 = 156 \text{ kips}$$

- Weight of reinforced concrete horizontal beam = 129 kips
  The number of piles at the abutment is

$$\frac{570 + 156 + 129}{100} \times 1.6 \approx 14 \text{ piles}$$

For two abutments 14 × 2 = 28 piles, where 1.6 is the approximate coefficient considering unfavorable conditions at work piles.

Alternative II

The amounts of materials for a span of 107.5 ft and its supports are taken from the Alternative I.

- Volume of reinforced concrete spans of 55 ft is:

    $1.0 \times 55 \times (26 + 2 \times 4.92) \times 6 = 55 \times 35.84 \times 6 = 11{,}827 \text{ ft}^3$

  A span of 107.5 ft is

  $$\frac{22446}{4} = 5610 \text{ ft}^3$$

  Total volume is $11{,}827 + 5610 = 17{,}437 \text{ ft}^3$
  Volume of precast concrete for intermediate spans:
- Supports for spans of 55 ft at the height above a foundation of 26 ft are $= 3529 \text{ ft}^3$
- Supports for a span of 107.5 ft volume (Alternative I)

  $$\frac{3239}{3} \times 2 = 2159 \text{ ft}^3$$

  Total volume $= 3529 + 2159 = 5688 \text{ ft}^3$
- Volume of reinforced concrete horizontal beams at intermediate supports for spans of 55 ft is $= 1637 \text{ ft}^3$. Horizontal beams for 107.5 ft spans are $= 953 \text{ ft}^3$
  Total volume is $1637 + 953 = 2590 \text{ ft}^3$
- Volume of reinforced concrete horizontal beams for abutments is 632 ft^3
- Volume of support slab for piles of intermediate supports for spans of 55 ft $= 3556 \text{ ft}^3$
  For spans of 107.5 ft = (Alternative I) $1658 \times 2 = 3316 \text{ ft}^3$
  Total volume $= 3556 + 3316 = 6872 \text{ ft}^3$
  To determine number of piles at the base of intermediate supports at spans of 55 ft, we calculate load on piles.
- Weight of reinforced concrete spans and surfacing:

    $(1.0 \times 1.148 \; 1.1 + 0.046 \times 1.5) \times 55.0 \; (26.0 + 2 \times 4.92)$

    $= 0.232 \times 55.0 \times 35.84 = 457 \text{ kips}$

- Live load and people on sidewalk:

  $$\text{Impact I} = l + \frac{50}{55 + 125} = 1.28$$

  Weight of the pier = 330 kips
  Number of piles at the base of the intermediate support:

  $$\frac{457 + 228 + 330}{100} \times 1.4 \approx 14 \text{ piles}$$

For four supports, 14 × 4 = 56 piles
Total of the intermediate supports: 56 + 29 × 2 = 114 piles
Load on piles under abutments under weight of span: 0.5 × 457 = 228.5 kips
Weight of live load and people on sidewalk:

$$(2 \times 0.64 \times 1.28 + 2 \times 0.06 \times 4.92) \times 1.4 \times 0.5 \times 55.0 + 2 \times 28.0 = 86 \times 56.0 = 142 \text{ kips}$$

Number of piles under abutment:

$$\frac{228.5 + 142 + 51.4}{100} \times 1.6 \approx 7 \text{ piles}$$

However, structurally it is not possible to accept less than two rows at six piles (considering the number of main girder at span).
For two abutments: 12 × 2 = 24 piles
Calculation of volume of material for Alternatives III and IV are made as at Alternatives I and II. We obtain

Alternative III

- Volume of reinforced concrete of spans is 19,461 ft^3
- Volume of assembled concrete of intermediate supports is 4806 ft^3
- Volume of reinforced horizontal beams at intermediate piers and abutments is 2876 ft^3
- Volume of slabs at piles for intermediate supports is 6101 ft^3
- Number of piles for intermediate supports and abutments = 130 piles

Alternative IV

- Volume of concrete for supports is 18,116 ft^3
- Volume of assembled concrete for intermediate support spans is 4806 ft^3
- Volume of reinforced horizontal beams at intermediate supports and abutments is 3239 ft^3
- Volume of slabs for piles is 2340 ft^3
- Number of piles is 105

Determination of Amounts of Materials for Alternatives. Calculations are shown in Table 14.1.

14.6.2 Example 1. Comparison of Alternatives of Bridge Structures

Owing to the preliminary nature of alternatives, they are compared on the basis of volume of materials used, also considering the general conditions in effect

TABLE 14.1 Example 1

Elements of Structure	Bridge Alternatives			
	I	II	III	IV
Prestressed concrete precast spans (ft^3)	22,446	17,437	19,461	18,103
Precast concrete supports (ft^3)	3,239	5,688	4,806	3,239
Massive reinforced concrete slabs above piles (ft^3)	4,975	6,872	6,101	4,976
Reinforced concrete horizontal beams of piers and abutments (ft^3)	1,644	2,590	2,876	2,340
Reinforced concrete piles (ft^3)	4,729	5,965	5,620	4,539
Total volume of materials (ft^3)	37,033	38,552	38,864	33,197
Total number of piles	109	138	130	105

during building and exploitation. Alternative IV uses a substantially smaller volume of materials than other alternatives. By using prestressed concrete, it is a little greater than Alternative II.

Alternative IV is also preferred from the point of view of building supports and small spans, which may be built faster using simpler means of transportation and erection equipment than spans over flood areas and other alternatives. Also, it is possible to note some exploitation preference for Alternative IV, and on this basis its structure was chosen. However, if a passing ice flow does not permit crossing at the location considered, by the application of thin pile supports at flood parts of the river, then it is necessary to use Alternative II, which also has a good parameter of volume of materials used.

14.6.3 Example 2. Comparison of Bridge Alternatives Across a Large River with a Navigable Clearance Crossed by a Steel Span (Figs. 14.3 and 14.4)

Elevation at the crossing axis, data regarding water level, and geological data are shown on Figure 14.4. The bridge should have two shipping spans with width clearances of 260 and 197 ft at high water and under the bridge a vertical clearance of 33 ft.

The bridge opening is 738 ft, the width of the roadway is 23 ft, and two sidewalks each have a width of 5 ft.

Appraisal of Local Conditions and Solution of General Problems of Construction. The transverse profile of a river crossing indicates that the river depth varies. However, generally, this asymmetry is relatively small, and we may assume that with time, depths may relocate in the direction of the right shore. For this reason, during design of the structure, it is advisable to consider symmetrical as well as asymmetrical schemes of alternatives. From the geo-

14.6 EXAMPLES OF BRIDGE ALTERNATIVES 271

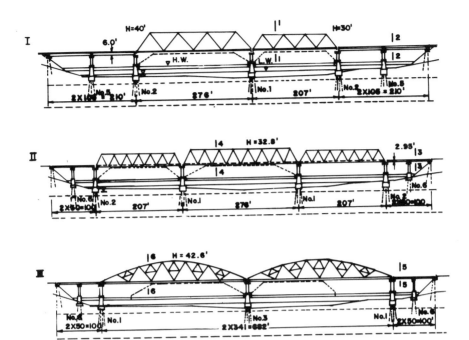

Figure 14.3 Schemes of Alternatives I, II, and III, for steel bridges.

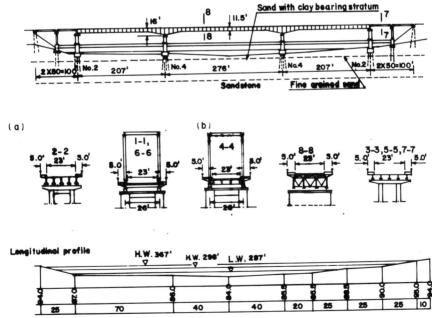

Figure 14.4 Scheme of Alternative IV for steel bridge; cross-sections and elevation.

logical profile it can be seen that a safe ground regarding settlement is located at a depth of 66 ft from the level of low water. Sands with layers of clay are located above this elevation. At the given geological conditions two solutions are possible:

- Arrangement of the foundation on friction piles having lengths of 33–40 ft
- Arrangement of the foundation reaching to the sandstone; in this case construction may be on piles, with shells having a diameter of 2.60–4.90 ft, which are driven into the weak grounds with the help of vibrodrivers and are bored another 6–10 ft into the sandstone. Piles for the abutments may be either reinforced concrete or shells driven by vibrodrivers.

Geological conditions at the crossing permit the construction of simple or continuous spans. With continuous spans, supports should be founded on piles reaching to sandstone, which prevents the settling of supports. The same bottom elevations of foundation slabs of channel piers are used because of the possible misplacement of maximum river depths along the river width. The bottom elevation of slab for high piles is 277 ft.

Starting from the water elevation it is possible to establish the following elevations for determining the dimensions of the structure:

- Elevation of the top of the foundation for channel piers: 288.64 − 1.64 = 287.00 ft, where 288.64 is the level of the low water.
- Elevation from which piers may have a lighter construction: 305.04 + 1.64 = 306.68 ft. It is assumed that the top of streamline part of the pier is at 1.64 ft higher than the level of high water.
- Elevation of the top level of shipping clearance: 298.48 + 32.80 = 331.28 ft.
- Elevation of the roadway limited by the shipping spans of the bridge: on spans having a low deck, using the construction height of approximately 4.90 ft; 331.28 + 4.90 = 336.18 ft.

During the development of alternative schemes consider the possibility of displacement of maximum river depths along the width of the river. Consider also schemes of alternatives with additional shipping spans and shipping spans with increased dimensions.

Detailed appraisal of basic problem data may illuminate possible alternatives. The characteristics of such alternatives, with an explanation of the system used and the dimensions of spans and supports are given below.

Design of Alternative Structural Schemes. Various possible alternatives of bridge structure are shown in Figures 14.3 and 14.4. In most alternatives shipping spans with low decks are used to obtain the lowest possible height for approach embankments. For comparison, also consider alternatives with an upper deck.

Alternative I. In this alternative measures are taken to obtain the minimum use of steel. For this reason, the number and dimensions of shipping spans used are minimal. Trusses of trapezoidal configuration with a rigid bottom chord are used as carrying structures, which is the most beneficial use of steel.

Spans not used for shipping are covered by reinforced concrete assembled structures. Such spans of 98 ft are used because: from one side, due to the great height of the bridge and the expense of constructing foundations, the number of supports should be smaller, and from the other side, it is not convenient to increase the dimensions of spans and thus increase the weight of the elements to be erected. Using spans with clearances of 98 ft permits both requirements to be successfully satisfied. A cantilever or portal crane may be used as the most convenient method for erecting spans.

Intermediate supports of shipping and nonshipping spans are built as monoliths. Abutments to the bridge are of lighter construction. Supports are based on highly reinforced concrete friction piles.

Alternative II. In this alternative three shipping spans are used because of the possible displacement of great depths of the river in transverse profile caused by erosion and deposition of sediment. In this case the cost and use of steel are increased by comparison with Alternative I, but safe shipping conditions are insured. Trusses of trapezoidal configuration with a flexible bottom chord and arrangement on the deck by transverse and longitudinal beams are used as carrying structures on the shipping spans. The amount of steel used in spans of this type is greater than on trusses with rigid bottom chords, but construction may be more convenient for fabrication and erection. For connection with embankment cones use assembled reinforced concrete spans that are 50 ft in length.

Reducing the dimension of reinforced concrete spans by comparison with Alternative I makes it possible to erect them using simple cranes. Also it is easier to transport prefabricated elements. Construction of piers for spans where $l = 50$ ft is as in Alternative I, with a corresponding reduction in dimensions. Abutments are of lighter construction when using piles. Supports for steel spans are the same as in Alternative I.

Alternative III. In this alternative two shipping spans are used. However, their dimensions are as large as 328 ft which insures, as in Alternative II, conditions of shipping during deformation of the channel and displacements of the larger depths of the river. Using this alternative, the number of supports is less by one than in Alternative II. Segmental trusses with rigid bottom chords are used as basic carrying structures over shipping spans, which at this large span are more convenient in construction and preparation than trusses of trapezoidal configuration. Intermediate supports are of the same construction as in Alternatives I and II, with foundations on friction reinforced concrete piles. Flood spans connect with embankment cones using the same method as in Alternative III.

Alternative IV. In this alternative where the construction of deck-type spans is considered to obtain the smallest possible construction height; composite beams are used. The advantage of small construction height is its relatively great flexibility, which permits the use of a continuous system and supports on a friction pile foundation. On continuous spans and more rigid trusses it will be necessary to use a more expensive foundation with supports on sandstone, for example, pile shells. The application of deck-type spans also allows us to reduce the dimensions of supports. Reinforced concrete spans are used at connections with the shore, as in Alternatives II and III.

Determination of Steel Weight in Steel Spans. It is assumed that steel spans are made from steel having resistances

$$\sigma_{all} = 30,000 \text{ psi}$$

The weight of the trusses is determined by Streletzkii's formula.[3] Figure 14.5 shows cross-sections of through trusses and plate girders of Alternative I and II.

Alternative I. Using a weight for the surfacing and reinforced concrete deck approximately equal to 0.111 k/ft^2, we obtain

$$g_{deck} = 0.111 \times 13 + 0.046 \times 5.75 + 0.010 \times 13 + 0.010 \times 5.75$$
$$= 1.443 + 0.264 + 0.130 + 0.058 = 1.895 \text{ k/ft}$$

where 0.010 k/ft^2 is the weight of steel in the deck structure. Live loadings are used for spans of 275 and 207 ft.

The corresponding dynamic coefficients are

$$1 + \frac{50}{275 + 125} = 1.125$$

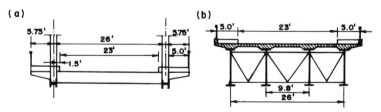

Figure 14.5 Schemes for the determination of coefficients of transverse load placement.

14.6 EXAMPLES OF BRIDGE ALTERNATIVES

$$1 + \frac{50}{207 + 125} = 1.150$$

The total value of equivalent loadings (considering the coefficient of transverse placement and dynamic coefficients) is

$$k_0 = 1.125\,(0.640 + 0.102) \times \frac{29.0}{26.0} + 0.06 \times 5.0 \times \frac{29.25}{26.0}$$

$$= 0.932 + 0.338 = 1.270 \text{ k/ft}$$

$$k_0 = 1.15\,(0.640 + 0.102) \times \frac{29.0}{26.0} + 0.06 \times 5.0 \times \frac{29.25}{26.0}$$

$$= 0.052 + 0.338 = 1.290 \text{ k/ft}$$

where 0.06 k/ft^2 is the load under the people. At a design resistance of 30,000 psi, the column weight is 0.49 k/ft^3 and the weight of bracings is $0.1\,g_{steel}$, taking values characteristics from Table 13.5 we obtain for spans of 275 and 207 ft:

$$g_{steel} = \frac{3.1 \times 1.4 \times 1.27 + 3.1 \times 1.2 \times 1.895}{\dfrac{50 \times 144}{0.49} - 3.1 \times 1.2 \times 275 \times 1.1} \times 275$$

$$= \frac{5.512 + 7.049}{8816.32 - 1125.30} \times 275 = \frac{12.561 \times 275}{7691.02} = 0.450 \text{ k/ft}$$

$$g_{steel} = \frac{3.1 \times 1.4 \times 1.29 + 3.1 \times 1.2 \times 1.895}{8816.32 - 3.1 \times 1.2 \times 207 \times 1.1} \times 207$$

$$= \frac{5.599 + 7.049}{8816.32 - 847.04} \times 207 = \frac{12.648 \times 207}{7969.276} = 0.329 \text{ k/ft}$$

The total quantity of steel consumed in the spans is the weight of steel in the trusses:

$$0.450 \times 2 \times 275 + 0.329 \times 2 \times 207 = 248 + 136 = 384 \text{ kips}$$

The weight of steel on the deck is

$$(0.010 \times 26.0 + 0.010 \times 5.74)(275 + 207)$$
$$= (0.26 + 0.115)(482) = 0.375 \times 482 = 181.75 \text{ kips}$$

The weight of steel in the bracing is $0.1 \times 384 = 38.40$ kips. The total quantity of steel consumed by Alternative I is 604 kips.

276 COMPARISON OF ALTERNATIVES

Alternative II. The weight of the deck, equivalent loading, coefficient of transverse placement, and dynamic coefficients are determined by calculating Alternative I. The weight of the deck supported by the longitudinal and transverse beams is $0.01\ k/\text{ft}^2$. The weight of the sidewalk support is $0.010\ k/\text{ft}^2$:

$$g = 0.111 \times 13 + 0.046 \times 5.75 + 0.015 \times 13 + 0.010 \times 5.75$$
$$= 1.443 + 0.265 + 0.195 + 0.057$$
$$= 1.960\ \text{k}/\text{ft}^2$$

Using Table 13.5, we obtain for spans of 275 and 207 ft:

$$g_{\text{steel}} = \frac{3.5 \times 1.4 \times 1.27 + 3.5 \times 1.2 \times 1.960}{\dfrac{30 \times 144}{0.49} - 3.5 \times 1.2 \times 2.75 \times 1.1} \times 275$$

$$= \frac{6.223 + 8.232}{8816.32 - 1270.5} \times 275 = \frac{14.455 \times 275}{7545} = 0.527\ \text{k/ft}$$

$$g_{\text{steel}} = \frac{3.5 \times 1.4 \times 1.290 + 3.5 \times 1.2 \times 1.960}{8816.32 - 3.5 \times 1.2 \times 207 \times 1.1} \times 207$$

$$= \frac{6.320 + 8.232}{8826.32 - 956.34} \times 207 = \frac{14.552 \times 207}{7859.98} = 0.383\ \text{k/ft}$$

Total quantity of steel used in three spans:
Weight of steel in trusses:

$$0.527 \times 2 \times 275 \times 2 \times 0.383 \times 207$$
$$= 289.85 + 158.85 + 158.56 = 449\ \text{kips}$$

Weight of steel at the deck:

$$(0.015 \times 26.00 + 0.010 \times 2 \times 5.75)(275 + 207)$$
$$= (0.390 + 0.115) \times (275 + 207)$$
$$= 0.505 \times 482 = 243.41\ \text{kips}$$

Weight of steel of the bracing:

$$0.1 \times 449 = 44.9 = 738\ \text{kips}$$

Alternative III. The weight of the deck (g_{deck}) and coefficients of transverse placement are taken from Alternative I. The equivalent loading for the span of 341 ft. The dynamic coefficient is

14.6 EXAMPLES OF BRIDGE ALTERNATIVES

$$1 + \mu = 1 + \frac{50}{341 + 125} = 1 + \frac{50}{466} = 1 + 0.11 = 1.11$$

$$k_0 = 1.11 \times (0.640 + 0.102)\frac{28.0}{26.0} + 0.06 \times 5.0 \times \frac{29.25}{26.0}$$

$$= 0.918 + 0.338 = 1.256 \ k/\text{ft}$$

$$g_{\text{steel}} = \frac{3.2 \times 1.4 \times 1.2256 + 3.2 \times 1.2 \times 1.895}{8816.32 - 3.2 \times 1.2 \times 341 \times 1.1} \times 341$$

$$= \frac{4.172 + 7.277}{8816.32 - 1440.38} \times 341 = \frac{11.499 \times 341}{7375.94} = 0.529 \ k/\text{ft}$$

Total quantity of steel consumed in two spans:
Weight of steel in trusses:

$$2 \times 0.529 \times 341 \times 2 = 721.56 \text{ kips}$$

Weight of steel for deck support:

$$(0.010 \times 26.0 + 0.010 \times 2 \times 5.75)\, 341 \times 2$$
$$= 0.375 \times 682 = 255.75 \text{ kips}$$

Weight of steel in bracing:

$$0.1 \times 721.56 = 72.16 \text{ kips}$$

The total quantity of steel consumed by Alternative III is 1050 kips.

Alternative IV. For one beam, when the weight of the deck is approximately $0.111 \ k/\text{ft}^2$ and of the sidewalk it is $0.046 \ k/\text{ft}^2$

$$g_{\text{deck}} = (0.111 \times 23.0 + 0.046 \times 2 \times 4.92)\frac{1}{4} = \frac{3.006}{4} = 0.752 \ k/\text{ft}$$

Coefficients for transverse distribution were determined above. Equivalent loading and dynamic coefficients were determined for spans of 275 and 207 ft for Alternative I.

For spans of 275 and 207 ft:

$$k_0 = 1.125 \times (0.640 + 0.102)\frac{29.0}{26.0} \times 0.62 + 0.06 \times 5.0 \times \frac{29.25}{26.0}$$

$$= 0.577 + 0.338 = 0.915 \ k/\text{ft}$$

278 COMPARISON OF ALTERNATIVES

$$k_0 = 1.15 \times (0.640 + 0.102) \frac{29.0}{26.0} \times 0.62 + 0.06 \times 5.0 \times \frac{29.25}{26.0}$$

$$= 0.590 + 0.338 = 0.928 \text{ k/ft}$$

From the weight characteristics in Table 13.4 and letting $g_{bracing} = 0.12 g_{steel}$, we obtain for spans of 270 and 207 ft:

$$g_{steel} = \frac{4.8 \times 1.4 \times 0.915 + 3.7 \times 1.2 \times 0.752}{8816.32 - 3.7 \times 1.2 \times 275 \times 1.12} \times 275$$

$$= \frac{6.149 + 3.334}{8816.32 - 1368} \times 275 = \frac{9.483 \times 275}{7448.32} = 0.350 \text{ k/ft}$$

$$g_{steel} = \frac{4.8 \times 1.4 \times 0.928 + 3.7 \times 1.2 \times 0.752}{8816.32 - 3.7 \times 1.2 \times 207 \times 1.12} \times 207$$

$$= \frac{6.236 + 3.334}{8816.32 - 1029.37} \times 207 = \frac{9.570 \times 207}{7786.95} = 0.255 \text{ k/ft}$$

Total quantity of steel consumed for whole bridge:
Weight of steel in the beams:

$$0.350 \times 4 \times 275 + 0.255 \times 4 \times 207 \times 2$$
$$= 385.00 + 422.28 = 807.28 \text{ kips}$$

Weight of steel in the bracing:

$$0.12 \times 807.28 = 96.87 \text{ kips}$$

The total quantity of steel consumed by Alternative IV is 904 kips.

Determining the Volume of Reinforced Concrete in Spans

Alternative I

- Volume of reinforced concrete in the slab of the deck and sidewalks of steel spans; thickness of the average slab in use = 6 in.

$$\tfrac{6}{12}(26.00 + 2 \times 5.75)(275 + 207)$$
$$= \tfrac{1}{2}(26.00 + 11.50)(482) = 9038 \text{ ft}^3$$

- Volume of reinforced concrete at assembled spans with prestressed reinforcing:

$$1.44(23.00 + 2 \times 4.92) \times 105.00 \times 4$$
$$= 1.44 \times 32.84 \times 105 \times 4 = 19{,}862.88 \text{ ft}^3$$

Alternative II

- Volume of reinforced concrete slabs of the deck and sidewalks:

$$\tfrac{6}{12}(26.00 + 2 \times 5.75)(275 + 2 \times 207)$$
$$= \tfrac{1}{2}(37.50 \times 689) = 12{,}919 \text{ ft}^3$$

- Volume of reinforced concrete assembled spans:

$$1.12(23.00 + 2 \times 4.92) \times 50.00 \times 4$$
$$= 1.12 \times 32.84 \times 200 = 7356 \text{ ft}^3$$

Alternative III

- Value of reinforced concrete slabs of the deck and sidewalks:

$$\tfrac{6}{12}(26.00 + 2 \times 5.75) \times 341 \times 2$$
$$= 37.50 \times 341 = 12{,}7888 \text{ ft}^3$$

- Volume of reinforced concrete spans (see Alternative II) is equal to 7356 ft^3

Alternative IV

- Volume of reinforced concrete slab of the deck and sidewalk:

$$\tfrac{6}{12}(23.00 + 2 \times 4.92) \times (275 + 2 \times 207)$$
$$= \tfrac{1}{2} \times 32.84 \times 689 = 11{,}313 \text{ ft}^3$$

- Volume of reinforced concrete spans (see Alternative II) is equal 7356 ft^3

Calculating the Quantity of Works at Supports. During the determination of the volume of supports, the height of the upper part H is used, considering that the top elevation of the support is lower by 3 ft than the upper limit of the clearance of the bridge, and that the bottom of the column is 1.50 ft higher than the level of high water. Therefore, $H_1 = (298 + 32 - 3) - (305 + 1.5) = 20.5$ ft.

The elevation of the foundation used is 1.5 ft lower than the level of low water. Thus the height of the bottom part of the pier is $H_2 = (305 + 1.50) - (288 - 1.50) - 20$ ft. The heights of the foundations used are equal to 7 and 10 ft.

When calculating the volume of works, we consider the six types of piers, shown in Figures 14.3 and 14.4. The volumes of piers under steel spans are determined by data shown in Table 14.2. The volumes of the works for piers

TABLE 14.2 Example 2

Pier Number	Volume of Reinforced Concrete—Upper Part of the Pier (ft³)	Volume of the Massive Part of the Pier (ft³)	Volume of Grillage and Piles (ft³)
1	3,705	6,705	8.293
2	2,928	5,010	8,327
3	5,010	8,998	12,526
4	—	9,809	8,293

5 and 6 under reinforced concrete spans and for pile foundations are estimated as follows. The volume of the reinforced concrete upper part is:

$$2 \times \frac{\pi \times 3.0^2}{4} \times 18.0 + 5.0 \times 4.5 \times 2.6 = 254.34 + 585.00 \approx 840 \text{ ft}^3$$

where 26.24 is the average length of horizontal beam. The massive bottom part is $6.23 \times 19.68 \times 22.96 = 2815$ ft³ where 26 ft is the average length of horizontal beam and $6.0 \times 20.0 \times 23.0 = 2760$ ft³ where 23.0 ft is the length of the pier. For the grillage and piles,

$$32 \times 1.3 \times 1.3 \times 46.0 + 36.0 \times 13.0 \times 9.0 = 2488 + 4212 = 6700 \text{ ft}^3$$

For pier 6, the volume of reinforced concrete upper part is

$$2 \times \frac{\pi \times 3.0^2}{4} \times 20 + 3.6 \times 4.6 \times 26 = 283 + 431 = 714 \text{ ft}^3$$

The massive bottom part is $5.0 \times 20.0 \times 23.0 = 2300$ ft³ and the grillage and piles are

$$15 \times 1.3 \times 1.3 \times 46.0 + 30 \times 13.8 = 1166 + 3120 = 4286 \text{ ft}^3$$

The amount of material used for abutments supported by piles is from the data of Example 1. The abutments for 105-ft spans are from Example 1, Alternative I. The volume of the horizontal beams is 1659 ft³ and 32 piles. For two abutments the volume of the horizontal beams is 608 ft³ and 24 piles. The general quantities of works for supports are given in Table 14.3.

Additional Earthwork at Approaches. In Alternatives I–III shipping clearances are covered by spans with decks at the bottom chords, whose construction heights had small differences between them. Alternative IV represents a deck span structure, the construction height of which is greater by approxi-

TABLE 14.3 Total Quantity of Works at Supports

Alternative Number	Reinforced Concrete Part of Supports (ft³)	Massive Parts Higher than Level of Foundation (ft³)	Reinforced Concrete Pile Foundation (ft³)
I	3705 + 2 × 2928 +2 × 937 + 1658.48 = 13,094	6704 + 2 × 5010 + 2 × 2815 = 22,354	8292 + 2 × 8327 + 2 × 6775 + 32 × 1.15 × 1.15 × 12 = 39,004
II	2 × 3705 + 2 × 2928 +2 × 766 + 365 = 15,433	2 × 6704 + 2 × 5010 + 2 × 2372 = 28,172	2 × 8292 + 8327 + 2 × 4605 + 24 × 1.15 × 1.15 × 12 = 34,502
III	2 × 3705 + 5010 + 2 × 766 + 635 = 14,587	2 × 6704 + 8998 + 2 × 2372 = 27,150	2 × 8298 + 12,526 + 2 × 4605 + 35 = 38,355
IV	2 × 2928 + 2 × 766 + 635 = 8023	2 × 5010 + 2 × 9809 + 2 × 2372 = 34,382	2 × 8237 + 2 × 8292 + 2 × 4605 + 35 = 43,483

mately 8.2 ft by comparison with the other alternatives. The embankments of approaches for this alternative will be higher by the same amount.

The height of the embankment on left shore at the location of connection with the bridge in Alternative IV is 34.44 ft and at the right shore it is 41 ft. The slopes of the locations are 25% and 10%. Considering that the slopes of the approaches are 4%, we obtain the following length for the approach embankments:

$$\text{Left Bank } \frac{3444}{29} = 118 \text{ ft at the average height of } 17.22 \text{ ft}$$

$$\text{Right bank } \frac{3444}{14} = 295 \text{ ft at the average height of } 20.5 \text{ ft}$$

The width of the embankments at the bottom of these heights is

$$47.32 + 2 \times 1.5 \times 17.22 = 47.32 + 51.66 = 98.98 \text{ ft}$$

and

$$472.32 + 2 \times 1.5 \times 20.5 = 47.32 + 61.50 = 108.82 \text{ ft}$$

Therefore, the additional volume of earthwork will be approximately

On left bank $98.98 \times 8.2 \times 118 = 95{,}773 \text{ ft}^3$
On right bank $108.82 \times 8.2 \times 295 = \underline{263{,}236} \text{ ft}^3$
Total $359{,}109 \text{ ft}^3$

14.6.4 Example 2. Comparison of Alternatives

The total quantity of basic works at the bridge alternatives and a comparison of them are shown in Table 14.4. Comparing the data given in Table 14.4 permits us to make the following conclusions:

1. *Alternative III*, which provides maximum standardization of construction and which has the advantage of allowing the displacement of maximum depths, requires a large amount of steel.

2. *Alternative II* has an advantage by comparison with Alternative I regarding the amount of steel used. However, this alternative has one more support, which without a doubt complicates the work. Alternatives I and IV are more economical, and the former requires substantially lower elevations at entrances to the bridge, which provide more convenient conditions for car traffic. However, Alternative I cannot insure normal shipping in the case of displacement of maximum depth transverse to the river.

TABLE 14.4 Example 2

Elements of Structure	Bridge Alternatives			
	I	II	III	IV
Steel spans (kips)	604	738	1,050	904
Reinforced concrete slabs of the deck and sidewalks (ft^3)	9,038	12,919	12,788	11,314
Reinforced concrete spans (ft^3)	20,689	7,209	7,209	7,209
Massive supports above the level of foundation (ft^3)	22,245	28,030	27,008	34,238
Reinforced concrete supports (ft^3)	12,060	15,302	14,456	7,892
Piles for supports (ft^3)	40,835	35,331	38,919	43,047
Embankments and approaches (additional earthwork)	—	—	—	359,108

14.6.5 Example 3. Design of Bridge Scheme Across a Large River

To illustrate a practical application of the basic principles discussed above for the design of a metal bridge, we consider the example of a metal bridge over a large river.[4] The highway crossing the shipping river requires one shipping span and has a clearance width under the bridge of 328 ft (100 m) and height of 41 ft (12.5 m). The width of the roadway on the bridge is 46 ft (14 m) and the width of each sidewalk is 5 ft (1.5 m). The live load used on roadways and sidewalks was according to the AASHTO specifications.

The geological conditions at the crossing location are as follows (Fig. 14.6). In the channel of the river are sands, supported by a layer of dense clay. At the right shore sands are changing to the sandy clay soil. The level of low water is −51.6 ft, the level of high water is 57.6 ft, and the shipping level −55.2 ft. The calculated velocity of the water at inundation is around 5.6 ft/sec. (1.7 m/sec). The required cross-section under the bridge is 6560 ft^2 (2000 m^2).

The shipping requirements in this case determine the positions of the bridge spans. At the middle part of the cross-section, where the maximum depths and main channel are located, it is necessary to locate the shipping span. The shore parts are covered by the shore spans. Considering the geological conditions, the foundations for the piers in the channel consist of reinforced concrete wells above the clay. For the shore supports, foundations consisting of reinforced concrete piles are used.

The first alternative utilizes a continuous three-span system deck system in which the main bridge system consists of plate girders. The bridge spans are 220 + 360 + 220 ft (67.0 + 110.0 + 67.0 m). In cross-section each span consists of six main plate girders. The reinforced concrete slab of the roadway performs as the composite section of the steel plate girders. The main plate girders have a constant height, which simplifies them and provides economy in their fabrication and erection. Close arrangement of the main girders is used

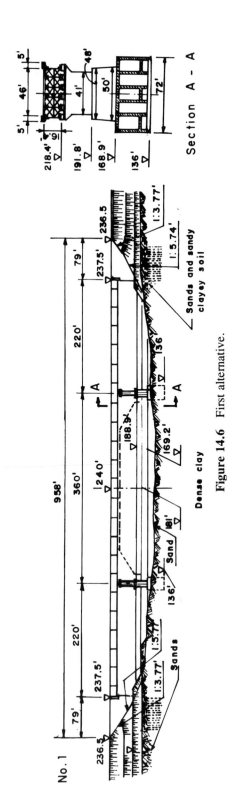

Figure 14.6 First alternative.

to eliminate cross beams and also to introduce a reinforced concrete slab into the composite work with the main girders.

The river piers of the bridge are massive concrete structures. To reduce the volume of the piers, their upper part has cantilevers in the cross section, which support the main edge girders of the span. The abutments of the bridge are reinforced concrete frames. The longitudinal slope of the bridge is 0.6%. The elevations of the approach embankments and the bridge ends are 72.1 ft.

The second alternative (Fig. 14.7) is different from the first because of the use of continuous deck trusses. Trusses have increased depth above river piers and parallel chords at the remaining length. The spans of the trusses are the same as in the first alternative. In the cross-section the span has four main trusses, spaced 16 ft apart (5 m). The deck of the bridge has a reinforced concrete slab consisting of the precast elements, which are placed on the transverse beams.

The supports of the bridge are similar in design to those of the first alternative. The bridge has two longitudinally sloping surfaces of 0.7%. The elevation at the top of the embankment approaches at the ends of the bridge is 73.1 ft. Therefore, the application of trusses instead of plate girders requires an increase in embankment height at approaches of 3.28 ft (1 m).

In the third alternative (Fig. 14.8) the shipping span is covered by a structure with a bottom chord, two main trusses with a polygonal upper chord, a triangular lattice, and a rigid bottom chord. Shore spans are covered by deck-type plate girders working as composite sections with the reinforced concrete slab of the deck. The spans in this alternative are 220 + 360 + 220 ft (67.0 + 110.0 + 67.0 m). The supports of the bridge are similar to the first two alternatives. Applying the shipping span trusses to the deck at the bottom chord to be lowered substantially allowed the height of the approach embankments. The elevations of the top of the embankments at the bridge ends are 70.0 ft.

The fourth alternative (Fig. 14.9) is like the third and has similar distribution of spans. A combined system in the shape of a flexible arch with a rigid tie and vertical suspenders covers the shipping span and gives a somewhat better external view than the truss with inclined diagonals used in Alternative III. The same spans, elevations of longitudinal profile, and supports are used as in the third alternative.

In the fifth alternative (Fig. 14.10) continuous three-span plate girder system is applied, which is reinforced at the middle span by a flexible arch. The arch at the supports is placed below the rigid tie beam, and at the middle part of the shipping span rises above the deck elevation. At the side spans there is a half-arch that extends from the river piers to the rigid tie beam at the section about 0.4 of the side span. The application of the half-arch and the continuity of the plate girder allow the height of the beam to be reduced somewhat by comparison with the third and fourth alternatives. As a result the heights of the approach embankments are somewhat lower. The arrangement of the supports permits a substantial reduction of the volume of material used for the river piers.

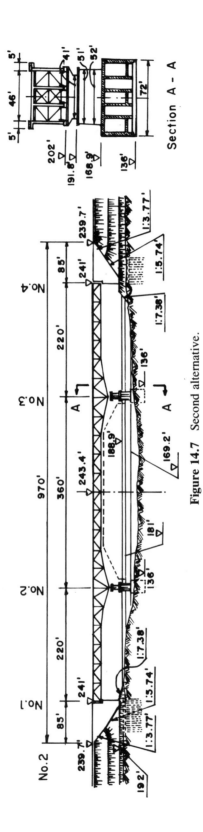

Figure 14.7 Second alternative.

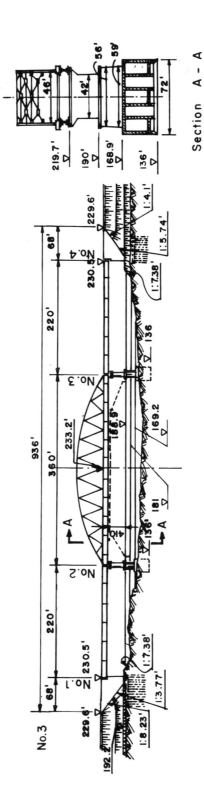

Figure 14.8 Third alternative.

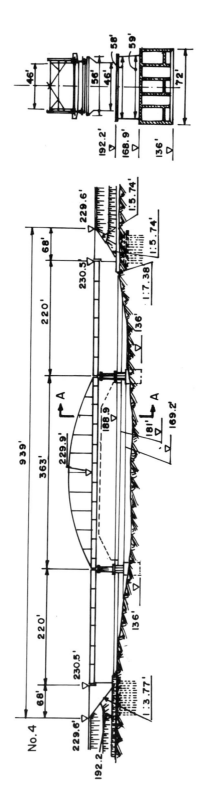

Figure 14.9 Fourth alternative.

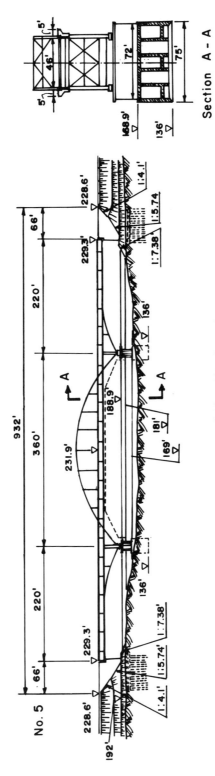

Figure 14.10 Fifth alternative.

For the most complete investigation of possible solutions, bridge alternatives with reinforced concrete spans are also considered. In the sixth alternative (Fig. 14.11) the shipping span is covered by a metal arch with two hinged deck-type crescent arches. The side spans are covered by the reinforced concrete two-hinged deck-type arches. The arrangement of the reinforced concrete side spans was undertaken to equalize the thrust of the middle and side spans and thereby eliminate large piers. Because of the presence of the thrust, the foundations of the supports are placed 1.6 ft (0.5 m) lower than the alternatives without thrusts. The arrangement of the reinforced concrete side spans reduces the amount of steel used on the bridge. Owing to the relatively small height of the deck, the elevations of the roadway on the bridge and the heights of the approach embankments in this alternative are the smallest.

The seventh alternative (Fig. 14.12) consists of a bridge on which all the spans are reinforced concrete. The spans are designed as a cantilever-beam system of prestressed concrete box sections. As a result the main beams are relatively low, allowing lower elevations of the approach embankments by comparison with the steel bridge (Alternative I). Connection of the bridge to the shores is achieved without abutments by the use of reinforced concrete cantilevers in the embankment. The river piers have foundations of reinforced concrete wells and the shore piers are supported by the piles. The volumes of basic works are shown in Table 14.5.

Comparing the alternatives is necessary to consider the economics and exploration criteria as well as the architectural qualities of each alternative. The most economical alternative is the seventh, which has reinforced concrete spans. The economical superiority of Alternative VII may be explained by the application of prestressed reinforced concrete structures for the spans. However it is impossible to accept Alternative VII because considering shipping requirements it is inconvenient during summer to block up the river by scaffolding. Erecting a superstructure using the cantilever method or concreting in the winter is relatively complicated and will be costly.

The alternatives that have shipping span trusses with the deck at the bottom chord (Alternatives III–VI) have minimal construction height, which results in the least expensive approaches. The cost of the bridge and the use of steel cause these alternatives to differ a little from the bridge alternatives having the deck at the upper chord.

The exception is Alternative VI in which, owing to the application of reinforced concrete side spans, a substantially smaller amount of steel is required. Regarding erection, the alternatives that have a roadway at the bottom chord are somewhat more complicated that bridges having the deck at the upper chord.

Regarding maintenance of the bridge, the deck at the bottom chord is less convenient because part of the basic carrying structure is not covered by the deck slab. Projecting elements of the top chord (or arches) with bracings above the roadway makes the external view of the bridge worse and lessens the ability of drivers on the bridge to observe the surrounding landscape. The external view of elevated bridge spans with the deck at the bottom chord is also some-

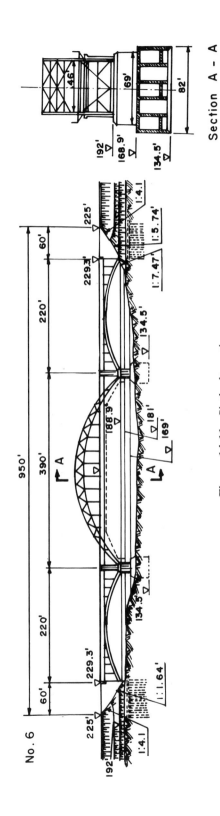

Figure 14.11 Sixth alternative.

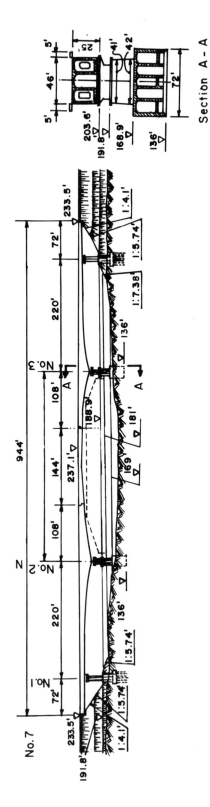

Figure 14.12 Seventh alternative.

TABLE 14.5 Volumes of Basic Works

Alternative	Reinforced Concrete Foundations (ft³)	Concrete for the Foundation (ft³)	Supports Reinforced Concrete Above Water (ft³)	Concrete above Water (ft³)	Reinforced Concrete Piles	Steel (tons)	Spans Reinforced Concrete (ft³)	Reinforced Concrete Deck (ft³)	Bearings (tons)
I	54684	77616	49392	84672	420	2500	—	38808	43
II	54684	77616	51861	81144	440	1920	—	35280	56
III	54684	77616	43394	95256	360	2270	—	28224	39
IV	54684	77616	43394	95256	360	2310	—	28224	39
V	55742	82202	44453	81144	330	2050	—	27166	50
VI	70560	123480	41278	148176	420	910	1420	15876	88
VII	49392	70560	41278	95256	256	—	2900	—	73

what less appealing than the architectural outline of a bridge with the deck at the top chord.

Regarding Alternatives I and II, which have deck-type steel girders, it is necessary to note that they require high embankments. This affects the cost of the approaches and may be inconvenient considering the general character of the longitudinal profile of the crossing. However, alternatives in which the bridge has the deck at the upper chord have a number of substantial advantages. The construction of continuous girder spans in Alternatives I and II is simple and convenient for fabrication and erection. The main girders may be easily erected by the longitudinal movement or by cantilevering. The deck at the upper chord protects the elements of the bridge against atmospheric action. While driving on the bridge one cannot see anything projected above the roadway elements of the structure and the surrounding landscape is able to be seen by the driver. The external view of the bridge is also better than when the deck is at the bottom chord.

By comparing the costs of the bridge alternatives with decks either at the top or at the bottom chord, we may conclude that Alternative I with continuous plate girders is the most expensive considering the cost of the bridge as well as the cost of the whole crossing with approaches. Also it requires the greatest amount of steel material. For this reason, irrespective of the construction advantages of plate girder spans and their nice external appearance, we should reject Alternative I.

Although Alternative II has a less attractive external view than Alternative I, it possesses all the advantages provided by having the deck at the upper chord and is similar in cost to the steel bridge with the deck at the bottom chord (Alternatives III–V). The cost of the bridge as well as the amount of steel material for Alternative II is smaller than for Alternatives III–V (Table 14.5). The general cost of the crossing (with approaches) for them is approximately equal.

TABLE 14.6 Volumes of Approaches and Total Costs of the Alternatives

Alternative	Approaches and River Regulation Structures			Total Cost of the Bridge in Millions of Dollars	
	Earth Works Approaches (ft^3)	Earth Works River (ft^3)	Reinforcing of Earth Works (ft^3)	Without Approaches	With Approaches
I	12,896,000	1,587,600	225,792	12.1	14.8
II	12,700,800	1,622,880	232,848	10.3	13.4
III	8,114,400	1,270,080	204,624	11.36	13.3
IV	8,114,440	1,270,080	204,624	11.36	11.36
V	7,761,600	1,217,160	197,568	10.86	12.760
VI	6,526,800	1,058,400	186,984	10.46	12.060
VII	9,702,000	1,340,640	239,904	8.5	10.9

Alternative VI with reinforced concrete shore spans has advantages in both total cost and necessary spending of steel by comparison with Alternative II. However, considering the positive characteristics of the deck at the upper chord and taking into account approximately the same cost for both alternatives, alternative 2 is recommended.

The considerations described above for comparing alternatives should be connected to the particular conditions (architectural requirements of the town, existence of construction materials, working conditions, etc.) of each case of bridge design. Consideration of these conditions may lead in different cases to different conclusions.

REFERENCES

1. Waddell, J. A. L., *Bridge Engineering*, vol. II, Wiley, New York, 1916, pp. 1182–1209.
2. Holgate, A., *The Art in Structural Design*, Clarendon Press, Oxford, 1986.
3. Polivanov, N. I., *Design and Analysis of Reinforced Concrete and Metal Highway Bridges*, Transport, Moscow, 1970, pp. 57–58 (in Russian).
4. Gibshman, E. E., *Metal Bridges on Highways*, 3rd ed., Avtotransizdat, Moscow, 1954, pp. 322–330 (in Russian).

CHAPTER 15

COMPUTER-AIDED DESIGN OF BRIDGES

15.1 PREPARATION OF THE GENERAL DATA

The first stage of the work for the bridge project preparation is choice of location, scheme and bridge system, and basic parameters of the construction considering the whole complex of local conditions. The application of computers at this stage of design in principle is possible. However, considering the complications of the problems that must be solved, computers are seldom used in this stage. Computers are usually applied at the second stage of design, when it is necessary to accomplish detailed calculation and establish the final dimensions of the structure. Computers are best used when they give substantial savings in time, especially when preparing complicated calculations of the same type.

During the design of bridges the analysis and calculations include the following:

a. Determination of the geometric characteristics of the sections of elements of construction (areas, resistance moments and inertia, sectorial characteristics of thin-walled sections, etc.) using given dimensions.
b. Determination of the coefficients of unknowns, determination of the free terms and solution systems of linear algebraic equations during the calculation of statically indeterminate systems.
c. Composition and solution of a system of differential equations during dynamic calculation of the bridges.
d. Calculation of the bridge considering space performance of the span structure.

e. Choice of cross-sections of construction. Usually points a–d take most of the time and calculation itself not as much time. For this reason, precise preparation of the initial data and good knowledge by the design engineer of the requirements and work order is very important in order to prepare the program.

The greatest affect is achieved when the design engineer himself performs calculations on computer and introduces, if necessary, design corrections of the initial data.

15.2 PLANNING AND BRIDGE DESIGN

During the last decade computers were widely used during the planning and design of bridges. In bridge planning and design computers are applied in automation of the design process. During computerized design the effect of the work is increased a few times by comparison with the manual design. The main effect is an increase in the quality of the structure under design resulting from the comparison of a greater number of alternatives, which enable the designer to choose the best one. During a comparison of the alternatives of bridge crossings by typical constructions, different schemes and lengths of the bridges, construction of supports and their foundations, standard girders, and location of the prefabrication plants are considered.

The criterion of optimization may be used for estimating the construction cost of the bridge crossing. The application of the computer to design elements of the bridge crossing as well as separate elements of the bridge (span structures, supports, regulation structures, embankments) should be a prepared algorithm, which is the sequence of calculation and logical action with tables permitting the preparation of a program for the computer. For example, in the design of bridge crossing it is necessary to introduce into the program more than 500 initial data. These characterize profile and geological sections along the axis of alternatives of bridge crossings, properties of the grounds, characteristics of the river, requirements of the designer concerning construction of span structures and supports, data regarding standard span structures and supports, costs of materials, prefabricated elements, and building works. By analogy it is possible with the computer to develop separate constructions, especially in series of typical supports and span structures.

The application of computers in bridge design may be very effective in the following cases:

1. During design of standard elements, when even a relatively small effect of optimization may provide greater economy, due to the application of larger quantities of standard construction.
2. Design of bridge crossings by large bridges and overpasses with wide use of standard constructions and elements. Here the effect is achieved

because of repetition of the problems, which are solved with the help of the same program.
3. Design of special bridges or new complicated bridge constructions; when program is applied it provides a large economical effect, which is substantially higher than the expenses for the development of a special program for this purpose.

Design of bridge structures includes construction and design, which are accomplished successively.

15.3 COMPUTER APPLICATION

Computer applications in engineering design have had an enormous effect on the analysis and design process in general. Automating analysis and design procedures has relegated much of the computational burden to the machines, allowing the engineer more time to evaluate alternatives and assume a more creative design and decision-making role. It can be said that their effect has been revolutionary.

The application of the computer in engineering can be divided into three stages. First is the computation and analysis of various kinds of complicated structures. In this stage, a great deal of research was devoted to the development of efficient algorithms for numeric computations. Then Computer-Aided Design and Drafting was developed. During this period, many softwares capable of generating drawings of structures were written with the development of the computer graphics industry. However, the greater bulk of "CADD" was actually computer-assisted drafting with the ability of the operator to manipulate the drawing and then proceed with the refinement of the original part or mechanism.[1] There were no "designs" done by the system. The engineer would do the design work and tell the machine what to draw or change. The third stage was the realization of automating analysis and design. In this stage, programs for design, analysis, and drawing were integrated into one system so that some automated system was generated. It was, undoubtedly, a great leap in the technology of computer application in engineering.

In the course of applying the computer in engineering the development of databases played an important role. The realization that the success of computer-aided engineering design must depend on efficient and reliable processing of a large volume of data has led to the concept of centralized data pools (i.e., databases). Databases serve as repositories of global data managed by a database system. The centralized data is shared by a number of application programs and users. The users and application programs have a logical view of data in the database. The physical layout of data within the database is irrelevant from their point of view. The logical view is referred to as data structure. The users and application programs can retrieve, modify, update, or delete data from the database using the logical view. The distinction between physical storage and logical layout of data, commonly referred to as data independence, is central to the concept of a database. The centralized database concept, with its data independence and flexibility, is currently the only promising avenue through which computer automation can be extended to the field of engineering design where a number of different disciplines are involved and information must be transmitted across disciplines.[2]

The computer also has wide application in bridge engineering. Many design packages have been written. Most of them are interactive and highly modular, consisting of segments for input, analysis, design, cost display, and graphical display. The following are design packages of bridge engineering developed in recent years. One dynamic package for landscape architecture design is also listed here.

15.4 BRADD-2 SYSTEM[1,3,4]

The Bridge Automated Design and Drafting (BRADD) system was developed by Michael Baker, Jr., Inc. in January 1985. It has been changed and updated several times.

15.4.1 Function

BRADD-2 is an integrated software that combines and automates the design, analysis, and drafting steps for certain types of highway bridges. The system utilizes state-of-the-art computer-aided design and drafting (CADD) technology to improve design productivity and reduce the time required to produce a contract document. At present, BRADD-2 has the following capabilities:

1. Complete superstructure design, substructure design, and drafting of simple span bridges from 18 to 200 ft.
2. The system allows interactive input using a menu-driving system from an alphanumeric terminal.
3. Superstructure types include precast concrete channel beams, adjacent prestressed concrete box beams, spread prestressed concrete box and I-beams, steel-rolled beams, and steel plate girders.
4. Substructure types include stub, cantilever, and wall abutments on spread or pile footings. Abutments wing walls or U-wing walls can be flared from 15 to 75 degrees.
5. A wide variety of cross-sections can be handled including normal, superelevation transitions, or full superelevated sections with or without sidewalks.
6. The horizontal alignment can be tangent or curved and a vertical curve can be located anywhere along the bridge.
7. Quantity and cost estimated calculations are automatically generated. This allows comparison and value engineering studies to be performed before final design drawings are generated. During the planning phase

of a project, this allows more accurate quantity and cost estimates to be obtained for budgeting purposes.
8. The design drawings generated by the system are to the appropriate scale and show all details. The drawings are generated in graphic design files using Intergraph hardware and software.
9. The system can design and draft abutments separately by using the "user-defined superstructure" type. This allows BRADD to be used for abutments for continuous structures.

15.4.2 Hardware

To run BRADD-2 the following hardwares are required:

1. CPU—Intergraph VAX
2. Workstation—Intergraph Workstation for PC (Note: once generated, the design files can be reviewed or modified on a UNIX-based standalone workstation or PC computer)
3. Alphanumeric terminal—any type
4. Storage—with source code, 120,000 blocks; without source code, 30,000 blocks. Note: BRADD-2 design files must be generated on the CPU specified.

15.4.3 System Software

The BRADD-2 system consists of five major subsystems. The subsystems are: Input, Design, Reports, Generation, and Plotting.

1. The Input subsystem controls the interaction of the user with the other BRADD-2 subsystems. The Input subsystems gather input data from the user for input to the database.

2. The Design subsystem contains all of the programs for the design of each component of the bridge.

3. The Reports subsystem generates additional data as requested by the user to help the user evaluate the design.

4. The Generation subsystem generates graphics design files containing the design drawing in standard scales and showing proper orientation and skew.

5. The Plot subsystem generates plots from the graphics design files. This subsystem primarily consists of the standard Intergraph plotting package plus routines to automate the generation of the plots.

15.4.4 Operation

BRADD-2 is an interactive system. A designer, by interactive dialog with the computer, enters appropriate information, span length, deck width, elevations, loading type, and desired superstructure and substructure types, as prompted by an alphanumeric terminal. The software processes the input information and generates cost estimates for alternative designs. The software also determines the size and quantity of construction materials and the geometry of their configuration. The software analyzes the structure and repeats the analysis and design process until the most economical design is found. Details are drafted automatically and displayed in two dimensions on screen. The designer reviews the designs and can modify individual elements. When satisfied, the designer then instructs a plotter to print hard copies.

15.4.5 Output

When placed in operation, the BRADD-2 will output the following items on the user's request:

1. Documentation of input information
2. Analytical results of either superstructure or substructure or both
3. Quantity of construction materials and cost estimates
4. The scaled design drawings

15.5 NEW IMAGE SYSTEM[5]

Image (i.e., Image Manipulate and Graphic Enhancement) is a dynamic software that was developed to help designers show their clients how a finished design will look prior to construction. It differs from BRADD-2 in that the software doesn't do any structural computation and analysis. Compared with the BRADD-2, the Image system is concerned more with art design than with structural design.

15.5.1 Function

The Image system is used for architectural and landscape design, graphic design and business sales presentation, beauty enhancement, cosmetic surgery and dental modification, interior design, and clothing and textile design. There is a database library consisting of actual photographic images of design elements in the Image system. With the database and the strong graphic management power the system enables designers to design anything they can imagine. The system has the ability to run the popular programs in the design field. This includes management, estimating programs, and sophisticated CAD packages for doing blueprints.

15.5.2 Hardware

To run the Image system, the following hardware is needed:

1. 40/80 Mb IBM computer or compatible
2. Video camcorder

3. Color monitor
4. Stylus pen and digitizing tablet
5. Video cassette recorder
6. Digital scanner

15.5.3 Operation

Suppose you are doing landscape design. First, you can go out to the site with a portable camcorder and videotape the views of the object for which you are going to render an image. This could include different views of the house and surrounding grounds. The next step is to go back to your office and connect the camcorder to the design system. Play back the scenes that you have captured and select the preferred image on the color monitor and save in computer memory. Then, you can select design elements from the system's extensive database library and manipulate them using the stylus and tablet with the site image on the screen. You can easily "cut and paste" with the stylus and tablet in order to remove an unwanted element and add something you like until you obtain the most satisfying design.

15.5.4 Output

Once the design is finished, it can be reproduced as full color "prints," photographically scanned into different sizes and formats or directly onto a VHS videotape.

15.6 BDES SYSTEM[6]

BDES (Bridge Design Expert System) is an expert system that applies the ideas of artificial intelligence to the bridge design process. This package can design for structural steel and prestressed concrete girders. Although in a preliminary stage, BDES now only considers superstructures of short- to medium-span bridges; it has shown the potential for application to expert systems in bridge design.

The so-called expert systems are intelligent computer programs capable of solving practical problems that have heretofore been considered difficult enough to require human intelligence for their solution. Expert systems attempt to model the problem-solving expertise of a human expert within a particular field. This requires representing specific knowledge of an expert as well as general problem-solving strategies. In knowledge-based expert systems, the expert's knowledge is stored in the system's knowledge base. This is analogous to a database in a conventional problem. The problem-solving strategy involves drawing inferences and controlling the reasoning process. These strategies comprise the inference procedure of an expert system.

The knowledge base includes two different types of knowledge: factual and heuristic. Factual knowledge can usually be found in textbooks and other references and hence is common knowledge. Heuristic knowledge is mostly private knowledge that experts have gained through experience. This knowledge is characterized by rules of good judgment, rules of good guessing, rules of plausible reasoning, and rules of thumb. These rules model the decision expertise the expert uses to solve the problem. This heuristic knowledge is represented in the form of rules and is thus referred to as the "rules." Rules in BDES may be used to select the superstructure type, determine the girder spacing, or decide between a simple or continuous span design.

BDES was constructed to explore the applications of expert systems to the design of bridge superstructures. BDES can proceed through the entire design process. BDES is highly user interactive with graphic capabilities to aid in input and output. Graphic displays guide the user in inputting geometry. Graphic output displays various cross-sections to illustrate clearly the designs generated by BDES.

BDES begins bridge design by inputing bridge geometry, bridge function, environmental information, materials, loading, design methods, and other constraints and criteria. Then BDES begins making design decisions using the rules. BDES will select the best alternative according to the "least weight" design and make structural analyses to verify its adequacy. If the design satisfies all requirements, it will be recommended to the user to use the design. If the design does not meet required specifications, BDES will let the user know why the design was not acceptable. However, at this time BDES does not redesign. The next phase in developing BDES would thus be to incorporate redesign rules.

15.7 BDS—BRIDGE DESIGN SYSTEM[7,8]

The Bridge Design System is an AASHTO software package. The software uses a database management system for design of selected components, allowing an engineer to analyze and design a bridge within a single integrated package. The leading feature that makes the BDS differ from other existing bridge design software is that it allows an engineer to pass information from component to component, superstructure to superstructure, and between phases of the design process (analysis to design to rating).

Engineers can define bridge requirements, create databases, load preexisting databases, and transfer data back and forth from the data structure and application subsystems. Alternate design checks can quickly be created by changing only a few database entries. Like BRADD-2, BDS is modular. Its major component programs are geometries, load building, analysis modeling solver, loading response processing, and specification checking. Other components, which may be added as future enhancements, are interactive design, rating, secondary components design, and quality cost generation. The software is being developed for use on both IBM and Digital Equipment hardware.

15.8 GEOMATH SYSTEM[9]

The system GEOMATH is written in C language with assembly language supplementation for critical functions. It is a program that enables the user to perform most of the computations involved in both alignment and structure geometry work. GEOMATH can process horizontal alignments consisting of tangents, arcs and clothoid spirals, vertical alignments consisting of grades and parabolas, and cross-sections consisting of gradients and break points. It provides the user with data necessary to lay out piers and parallel and flared straight girder arrangements. It also computes beam seat locations and deck elevations.

15.9 APPLICATIONS OF MICROCOMPUTERS IN BRIDGE ENGINEERING[10]

Although investigation in the United States of the use of microcomputers in bridge design activities showed that the use of microcomputers was found to be limited, subsequent research indicated that the current generation of 16-bit machines offers significant advantages. It can be seen from the following that the microcomputer will get wide use in bridge design application.

1. Bridge designers who use mainframes as their primary computing resource felt the need to improve their computing environment. The reasons included slow turnaround time on time-shared systems, a desire for better access to software, and insufficient access to terminals connected to the mainframe.

2. The development of hardware and operating sytems of microcomputers makes it possible to run large bridge design applications with a microcomputer either as standalone or as an intelligent terminal linked to the mainframe.

3. Microcomputer software for bridge design applications can be gotten by downloading and converting the existing mainframe programs. The following converted mainframe programs are currently available and can be obtained by contacting the bridge division in the appropriate state:

    a. Prestressed Concrete I-Beam Design and Analysis (standard AASHTO and nonstandard simple-span bridge girders) (Virginia)

    b. Steel Bridge Girder Design and Analysis (Virginia)

    c. Deck Slab Design (Virginia)

    d. Critical Moments and Shears on a simple span (Virginia)

    e. Georgia Bent Program (South Dakota)

    f. Continuous Span Prestresses Concrete Bridge Girder Design (South Dakota)

    g. PCA Reinforced Concrete Column Design

There are several advantages to running large-scale converted mainframe bridge design software on a microcomputer. It provides greater flexibility to the engineer. Applications can be run at any time without the need for access to a mainframe. Microcomputers can insulate bridge designers from the inconveniences of unscheduled mainframe downtimes, and so on.

The development of microcomputers signals a new era of computer use. The significant computing power they possess, along with their relatively low cost compared with traditional large computers, has assured their success. Their use is being constantly explored in many business and engineering applications.

15.10 DCA STRUCTURAL ENGINEERING SOFTWARE

DCA's structural engineering software is a series of powerful programs created to automate the design and production of plans, details, and analysis models. These programs run completely within AutoCAD, providing an interactive graphic interface that allows the user to create, edit, view, and plot all types of drawings quickly and easily. The following are two programs of DCA's software.

15.10.1 STRUCTURAL DESIGNER

The most significant feature of STRUCTURAL DESIGNER is its strong drawing management techniques and its industry-specified materials libraries. By working within AutoCAD's graphic interface, users have instant access to the industry standard construction materials, and can use that information to interactively assemble building components for the creation of:

1. *Framing Plans.* Quickly draws girders, columns, joints, and all structural members to form floor, roof, mezzanine, and wind bent plans. Automates grid and structural frame layout.
2. *Foundation Plans.* Constructs entire foundation plans with footings, walls, plates, piers, pilasters, grade beams, control joints, boring locations, and contours. The STRUCTURAL DESIGNER draws the entire foundation system based on specified footing dimensions.
3. *Structural Concrete Plans.* Creates plans of concrete parking garage structures and high-rise office buildings with poured and precast materials; columns, slabs, tees, waffles, and hollow-core planks.
4. *Tilt-up Panels.* Automatic creation of plans and details for concrete tilt-up panel construction.
5. *Elevations, Sections, and Details.* Elevations, sections, and details are assembled with industry standard materials from the building components library. An effective means of creating and maintaining large libraries of standard and custom details.
6. *3D Wire Frame Diagrams.* Three-dimensional models are easily generated and provide valuable information for use in structural analysis and design.

7. *Render Ready Models.* Generate impressive presentation drawings with the STRUCTURAL DESIGNER, stick to 3D shape modeler and AutoSHADE.
8. *Full Engineering Construction Plans.* By combining the automating features of the STRUCTURAL DESIGNER with the powerful CAD capabilities of AutoCAD, users can take the project from schematic design all the way through the production of complete engineering construction plans, virtually eliminating manual drafting.

Utilizing the powerful design interface function of AutoCAD, analysis models of structures can be created and brought into STRUCTURAL DESIGNER as stick frame diagrams with members, loads, supports, and property sets. They can be automatically converted into initial framing plans, then completed as final drawings. Extremely efficient in its operation, this program link is invaluable for preparing input data, and eliminates the need to manually calculate the geometry of a model and prepare large batch input files.

15.10.2 STEEL DETAILER

The STEEL DETAILER is a truly advanced drafting system for the structural detailer. With a sophisticated Detail Member Manager, the program provides all the connectivity information for each member while working inside AutoCAD. The member manager eliminates the need to reference erection plans by providing exact information about member lengths, locations, and angles for all members.

The program's drafting tools are linked into this intelligent member database to parametrically generate each part of a drawing. Special tools are provided for beams, columns, bracing, stairs and handrails, erection drawings, anchor bolt plans, advance bill of material, field bolt generator, and dynamic dimensioning. The software comes with computerized catalogues of AISC, CISC, BS4, and DIN steel components. Bolts, nuts, bars, and plate specifications are integrated with the program.

To run the DCA Structural Engineering Software, one of the following hardwares may be chosen:

1. *Personal Computer.* IBM XT, AT, IBM PS/2, or 100% IBM compatible computer.
   a. AutoCAD Release 10, ADE-3
   b. MS-DOS version 3.0 or greater
   c. 20 MB hard disk
   d. 640 K RAM
   e. Math coprocessor
2. *Workstation*
   a. Sun Microsystems Sun-3 series with floating point coprocessor
   b. Sun Microsystems Sun 386i

c. Sun Microsystems SPARC stations
d. AutoCAD Release 10
e. SunOS Release 4.0

15.11 DIRECT OPTIMAL DESIGN OF CONTINUOUS COMPOSITE STEEL BRIDGE GIRDER[11]

The traditional design of a continuous steel highway bridge girder is produced by a process of iteration: a trial configuration is selected, checked for specification acceptability, and then reconfigured, in a cycle process that continues until the result is judged to be adequate.

There are two types of criteria to be considered in the design of any such structure. One involves a hard set of constraints that are drawn from the governing specification, none of which may be violated if the design is to conform to that specification. The second criterion represents the merit of the resulting design. It is somewhat softer because there actually is no specific requirement that the design actually is to be the most meritorious possible, although the conscientious designer certainly hopes at least to approach that state.

The prime objective of traditional design is to bring the configuration into compliance with the specification. While this is being accomplished, the objective of achieving high merit in the design tends to become secondary in precedence, though perhaps not completely ignored. As a result, the merit of the final design basically has been determined at the outset, when the design policy and preliminary design were established.

With the use of the computer, a design can be accomplished completely by mathematical methods. The problem simply is to find an optimal set of values for the design variables defining the configuration of the system, while at the same time taking care not to violate a set of constraining relationships, typified by stress and plate slenderness limitations. The merit of the system must be definable, such as weight or cost, and the constraining relationships can, in the case of highway girders, be assembled directly from the AASHTO specification. The formal mathematical representation of this more general problem is

$$\text{Minimize } F(x_j)$$

$$\text{Subject to } g_i(x_j) < 0 \tag{15.1}$$

where x_j represents the entire set of design variables, j ranges from 1 to the total number of such variables, and g_j represents the entire set of governing constraints, i ranges from 1 to the total number of such constraints.

The problem represented by Eq. (15.1) is well known as the mathematical programming (i.e., optimization) problem, and a number of solution techniques have been developed for its resolution. Only recently, however, have these techniques been able to be applied to such large design problems as highway girder design, and that only because of recent technological advances in computing hardware and software.

One of the methods is preferred here because it can be represented as a manipulation of matrices. If the mathematical model for analysis is viewed as

$$\begin{aligned} a_{11}x_1 + a_{12}x_2 + \cdots + a_{1n}x_n &= b_1 \\ a_{21}x_1 + a_{22}x_2 + \cdots + a_{2n}x_n &= b_2 \\ &\vdots \\ a_{n1}x_1 + a_{n2}x_2 + \cdots + a_{nn}x_n &= b_n \end{aligned} \qquad (15.2)$$

representing the relationships between stiffness and displacements for an articulated structure, convertible by other relationships to internal forces, then the process of design similarly might be viewed as being represented by one of the mathematical models capable of solving (15.1), which involves solving a sequence of linear programming problems, each LP programs in the sequence rendered as

$$\begin{aligned} c_1x_1 + c_2x_2 + \cdots + c_nx_n & \\ a_{11}x_1 + a_{12}x_2 + \cdots + a_{1n}x_n &= b_1 \\ a_{21}x_1 + a_{22}x_2 + \cdots + a_{2n}x_n &= b_2 \\ &\vdots \\ a_{m1}x_1 + a_{m2}x_2 + \cdots + a_{mn}x_n &= b_n \end{aligned} \qquad (15.3)$$

In the equations above, c_j is obtained by differentiating the cost function with respect to design variable x_j; and a_{ij} is obtained by differentiating constraint i with respect to variable j. In general, each of these coefficients, as well as the right hand of the constraint set, b_i, is a function of the design variables x_j.

The problem is solved by first postulating a trial design, in an interesting parallel to the traditional approach to the design, and using the values of design variables in the trial design to evaluate the coefficients c_j, a_{ij}, and b_i. The solution is fed back for a reevaluation of the coefficients, setting up yet another LP problem to be solved again. Do this in a cyclical manner, always feeding back the solution to obtain the coefficients for the next cycle, until the solution to the current LP problem is sufficiently close to the solution of its immediate prodecessor, at which point convergence has occurred and the nonlinear problem (15.1) has been solved.

15.12 MODEL FOR THE INTEGRATION OF DESIGN AND DRAFTING SOFTWARE[12]

A methodology for integrating application design programs and Computer Aided Design (CAD) packages has been developed. Application of the methodology enables the automatic translation of the final design information into production drawings of the design, reducing the possibility of translation errors and increasing the engineer's productivity. The use of the methodology can greatly reduce the time required to translate design information into production drawings. The drawing files were generated using the Initial Graphics Exchange Specification (IGES), a neutral drawing file data format that can be utilized by a variety of CAD systems.

The methodology described is intended to serve as a general model for accomplishing the link between design and drafting software. The model has been implemented in the microcomputer environment using a reinforced concrete box culvert design program: BRASS-CULVERT. A processor has been written that produces detailed drawings of the culvert design automatically. These drawings are generated in a data format that can be utilized by a wide variety of CAD systems. While the demonstration of this particular model is limited to reinforced concrete culvert design and drafting, the techniques described can be applied to a wide variety of engineering and engineering-related applications.

15.13 THE EFFECTIVE USE OF CADD IN BRIDGE DESIGN[13]

Most CADD libraries consist of the material supplied by the CADD vendor. The library is usually a collection, rather like a scrap book. Every column CADD drawn can be found there, as can every barrier, deck cross section, abutment, wall, and so on. The reason for saving every detail is that some time in the future the user will be able to browse through the hard copy library and select the appropriate detail in his application. Unfortunately, in practice the library will become so unwieldy that the CADD user will not be prepared to go through it. The details that will be used will be those used on the last job. This will lead to a second problem; the CADD detail will not be exactly what is needed and will require on-screen editing. Once on-screen editing is necessary, many of the advantages of a CADD library will be lost.

In a parameter library, details are not stored as multiple varieties to include every previous example created. Instead, commonly used details are stored as an X-BASIC routine, which will draw the detail from a series of dimensions supplied as parameters by the CADD user. These dimensions are developed by designers in the normal way and passed on to the CADD user. The CADD user runs the X-BASIC program, which is written in a question-and-answer format. The CADD user answers the prompted questions on the screen. Once all the parameters have been provided, the X-BASIC routine automatically generates the required detail, including all dimensions and leader lines, and places it in the plan sublibrary, ready for the CADD operator.

From demonstrations drawing a bridge-deck cross-section and bridge abutment, it can be seen that the CADD library, which contains every detail and variation ever used by an office, is actually counterproductive. A sparse library containing parametric routines capable of generating a unique drawing or detail when needed is more efficient and lends itself to the ultimate goal of linking design and CADD.

REFERENCES

1. Scott, W. M. and Pater, M. G., Development and utilization of the bridge automated design and drafting system, Fall 1989, Lecture Series, *The Structural Group, Boston Society of Civil Engineers Section, American Society of Civil Engineers.*
2. Humar, J. C., Case studies on the application of CAD in civil engineering, computer applications in structural engineering, *ASCE*, 1987.
3. Tiberio, T. J., Hayward, J. C., Mahendra G., Patel, M. G., and McClure, R. M., A bridge automated design and drafting system (BRADD-2), *Transportation Research Record* 1118.
4. Campbell, J. J., Flango, R. J., Martin, H., Rosick, M. P., and Smyers, W. L., The development of a bridge automated design and drafting system, *Proceedings, 5th Annual International Bridge Conference, Engineers' Society of Western Pennsylvania, 1988.*
5. *New Image Design System*, New Image Industries Inc., 21218 Vanomen St., Canoga Park, CA.
6. Welch, J. G. and Biswas, M., Application of expert systems in the design of bridges, *Transportation Research Record* 1072.
7. Campbell, J., Bridge on screen, *Civ. Eng.*, Sept. 1990.
8. Garson, R. C., Hayward, J. C., Gloyd, C. S., and Imbsen, R. A., The AASHTO bridge design system. A report of research in progress, *Proceedings, 5th Annual International Bridge Conference, Engineers' Society of Western Pennsylvania*, 1988.
9. *Geomath*, Leap Software Inc., P.O. Box 290414, Tampa, FL.
10. Love, R. A., Barton, F. W., and McKeel, W. T., Jr., Application of microcomputer in bridge design, *Transportation Research Record* 1072.
11. Douty, R. T., Direct optimal design of continuous composite steel bridge girder, *Proceedings, 5th Annual International Bridge Conference, Engineers' Society of Western Pennsylvania, 1988.*
12. Guenther, P. W. and Puckett, A., Model for the integration of design and drafting software, *Proceedings, 5th Annual International Bridge Conference, Engineers' Society of Western Pennsylvania*, 1988.
13. Luscombe, R. W., The effective use of CADD in bridge design, *Proceedings, 5th Annual International Bridge Conference, Engineers' Society of Western Pennsylvania, 1988.*

INDEX

Abutment(s), 159–165
  arch bridge
    superstructure (reinforced concrete bridge), 139
    superstructure (steel bridge), 114
  design analysis, 164
  generally, 159
  girder bridge, preliminary design methodology, 246–248
  material for, 163
  seismic design, 194
  structural system, superstructure and substructure, 86, 87
  types of, 159–163
    box-type abutment, 161–162
    flanking-span abutment, 162
    floating abutment, 163
    straight-wing abutment, 160–161
    U-type abutment, 161
    wing-type abutment, 159–160
Aesthetics, 166–176
  bridge location, 46, 167
  design methodology, 201, 206–208
  failures in, 170
  generally, 166
  requirements for, 167–170
  steel bridge, 170–175
Akaghi suspension bridge (Japan), 38–39
Alternative comparison, 261–295
  calculated cost method, 261
  examples, 264–295
    reinforced concrete bridge, 265–270
    steel bridge, 270–295
  external view, 264
  fabrication and erection conditions, 262–263
  generally, 261
  materials cost method, 261–262
  performance conditions, 264

preliminary design methodology
  basic work determination during design, 253
  sequence of work during, 227
American Association of State Highway and Transportation Officials (AASHTO), 98, 100, 131, 177, 180, 190, 191, 283, 303
Ammann, Othmar Herman, 25
Ancient period, bridge development in, 5–8
Angle of crossing. *See* Layout
Annacis Island cable-stayed bridge (Canada), 32
Approaches:
  crossing design
    regulation and bank protection structures, 68–69
    water passage data, 64
  earthwork at, steel bridge, alternative comparison example, 280–282
  layout, 59
Aqueducts, Roman period, 11
Arch bridge:
  aesthetics, 172
  history of, 8, 11–18, 21
  reinforced concrete bridge
    design methodology, theoretical methods of, 220
    preliminary design methodology, 236–237, 250–252
    superstructure, 138–141
  steel bridge, superstructure, 112–114
  structural system, bridge types, 95
Avignon Bridge (France), 14

Bank protection structures, crossing design principles, 68–69
Batter, pier, 152
Bayonne Bridge (U.S.), 21–22

311

Beam bridge:
  aesthetics, 171
  composite beam and girder bridge, superstructure (steel bridge), 99–100
  plate-girder bridge, superstructure (steel bridge), 98–99
  rolled-beam bridge, superstructure (steel bridge), 97–98
  structural system, bridge types, 95
Bearings, structural system, superstructure, 88
Belting course, pier, 151
Bending movement, load distribution, specifications and codes, 187–188
Benezet, Saint, 14
Bering Strait bridge (U.S.-Russia), 39–41
Box girder, orthotropic deck bridge, superstructure, 107
Box-girder bridge:
  composite, superstructure (steel bridge), 105–106
  superstructure (reinforced concrete bridge), 131–132
Box-type abutment, described, 161–162
BRADD (Bridge Automated Design and Drafting) system, 299–301
Bridge(s). *See also entries under specific types of bridges*
  future of, 36–42. *See also* Future bridges
  history of, 1–44. *See also* Historical bridges
Bridge Automated Design and Drafting (BRADD) system, 299–301
Bridge crossing design. *See* Crossing design
Bridge Design Expert system (BDES), 302–303
Bridge Design System (BDS), 303
Bridge geometry:
  arch bridge, superstructure (reinforced concrete bridge), 140
  structural system, 88–92
Bridge layout. *See* Layout
Bridge location. *See* Location
Bridge opening:
  crossing design
    computation of, scour and erosion, 81–84
    design of, 66–67
  preliminary design methodology, 231
  structural system, geometry, 88, 90
Bridge pad, pier, 151
Bridge Sant'Angelo (Italy), 11, 12
Bridge seat, pier, 151
Bridge structural system. *See* Structural system; Substructure; Superstructure
Bridge system, design methodology, basic parameters, 212
Britannia Bridge (U.K.), 17–18
Brooklyn Bridge (U.S.), 24–25*

Cable-stayed bridge:
  aesthetics, 173–175
  future of, 37–38
  history of, 30–32
  structural system, bridge types, 95
  superstructure (reinforced concrete bridge), 141–145
  superstructure (steel bridge)
    composite forms, 126
    deck types, 121
    generally, 118
    main girder and truss, 121–124
    stay cable arrangement, 118–119
    stay cable position, 119–120
    structural advantages, 124–125
    suspension bridge compared, 125
    tower types, 120–121
Cable-supported box girder, orthotropic deck bridge, 107
Cable system, suspension bridge, superstructure (steel bridge), 115–116
Calculated cost method, alternative comparison, 261
Cantilever-arch bridge, reinforced concrete bridge, preliminary design methodology, 237
Cantilever bridge:
  aesthetics, 171–172
  history of, 8, 20–21
  prestressed concrete bridge, superstructure (reinforced concrete bridge), 148–149
  reinforced concrete bridge, preliminary design methodology, 234
Cast-in-place balanced cantilever, prestressed concrete segmental bridge, 132–133
Cast-in-place prestressed concrete, prestressed concrete bridge, 148
Chords, structural system, superstructure and substructure, 87–88
Clearance:
  preliminary design methodology, 231–232
  structural system, geometry, 91
Clear span, structural system, geometry, 90–91
Coalbrookdale bridge (U.K.), 17
Codes. *See* Specifications and codes
Color, aesthetics and, 169
Column support, girder bridge, preliminary design methodology, 248
Comparison of alternatives. *See* Alternative comparison
Composite bridge:
  beam and girder bridge, superstructure (steel bridge), 99–100
  box-girder bridge, superstructure (steel bridge), 105–106

INDEX   313

preliminary design methodology, 237–246
weight of steel, 256–259
Computer-aided design, 296–310
　applications, 298–299
　Bridge Automated Design and Drafting (BRADD) system, 299–301
　Bridge Design Expert system (BDES), 302–303
　Bridge Design System (BDS), 303
　continuous composite steel bridge girder, direct optimal design, 307–308
　data preparation, 296–297
　DCA structural engineering software, 305
　GEOMATH system, 304
　Image system, 301–302
　integration of design and drafting software model, 308–309
　microcomputer applications in, 304–305
　planning and design, 297–298
　STEEL DETAILER, 306–307
　STRUCTURAL DESIGNER, 305–306
　use of, 309–310
Concrete, reinforced concrete bridge, precast simple-span, 253–256. *See also* Reinforced concrete; Reinforced concrete bridge
Concrete bridge (historical), 32–35
　prestressed concrete, 34–35
　reinforced concrete, 32–34
Continuity, suspension bridge, superstructure (steel bridge), 117
Continuous composite steel girder bridge:
　direct optimal design, computer-aided design, 307–308
　superstructure (steel bridge), 100–103
Contrast, aesthetics and, 169
Coping, pier, 151
Coppelen Memorial Bridge (U.S.), 33
Crossing design, 62–85
　layout conditions, 69–73
　openings, computation of, 73–76
　　definitions, 73–74
　　procedures, 74–76
　principles, 63–69
　　approaches, regulation and bank protection structures, 68–69
　　exploration stage, 65–66
　　location selection and bridge opening design, 66–67
　　scheme and sequence in, 67–68
　　water passage data, 63–65
　scour, 76–84
　　bridge opening computation and, 81–84
　　estimate of, 81
　　factors affecting, 77–79
　　foundation protection for, 79–80

　　generally, 76
　　local scour, 76–77
　　minimization of effects, 80–81
　span layout, 62–63

DCA structural engineering software, 305
Dead loads, specifications and codes, loads on bridge, 179–180
Deck-girder bridge, superstructure (reinforced concrete bridge), 129–131
Deck-type truss, preliminary design methodology, 240
Design aesthetics. *See* Aesthetics
Design methodology, 198–223. *See also* Preliminary design methodology
　assumptions in, 201–210
　　aesthetic requirements, 206–208
　　basic requirements, 204–205
　　generally, 201–204
　　scientific research requirement, 208–210
　basic parameters, 210–214
　　bridge system, 212
　　generally, 210–212
　　span construction type, 213–214
　　span size, 213
　　support types, 214
　computer-aided design, 296–310. *See also* Computer-aided design
　theoretical methods of preliminary design, 214–223
　　final alternative selection (reinforced concrete bridge), 219–222
　　generally, 214–218
　　practical methods, 218–219
　trend characteristics, 198–201
　　creative trend, 200–201
　　generally, 198–199
　　practical trend, 201
　　rational computation trend, 199–200
Drafting software, integration of design and, 308–309

Eads, James, 18
Eads bridge (St. Louis bridge, U.S.), 18–20
Earth pressure, specifications and codes, 190
Earthquake. *See* Seismic design
Earthwork, steel bridge, alternative comparison example, 280–282
Economic principles:
　aesthetics and, 167
　design methodology assumptions, aesthetic requirements, 207–208
　prestressed concrete bridge, superstructure (reinforced concrete bridge), 148
　span determination, structural system, 92–95

Erosion, crossing design, bridge opening computation and, 81–84. *See also* Scour
Esla bridge (Spain), 33
Exploitation requirements, design methodology assumptions, 205
Expressiveness, aesthetics and, 168
External view, alternative comparison by, 264

Field work, layout, site investigation, 60
Finsterwalder, first name, 35
Firth of Forth bridge (U.K.), 20
Fixed arch:
  reinforced concrete bridge superstructure, 138
  steel bridge superstructure, 113
Flanking-span abutment, described, 162
Floating abutment, described, 163
Floor beam, orthotropic deck bridge, 108–109
Floor system, suspension bridge, 116–117
Footing course, pier, 151–152
Foundation(s):
  design of
    basic parameters, 212
    seismic design, 194
  protection of, for scour, crossing design, 79–80
  site investigation, field work, 60
Frame bridge:
  preliminary design methodology, reinforced concrete bridge, 233–234
  superstructure, reinforced concrete bridge, 137
Freyssinet, Eugne, 34
Future bridges, 36–42
  Akaghi suspension bridge, 38–39
  Bering Strait bridge, 39–41
  Gibraltar Strait bridge, 41–42
  Messina Straits suspension bridge, 36–37
  Normandy cable-stayed bridge, 37–38

Geological data, preliminary design methodology, 228–230
GEOMATH system, 304
Geometry:
  arch bridge, superstructure, reinforced concrete bridge, 140
  structural system, 88–92
George Washington bridge (U.S.), 25–26
Gibraltar Strait bridge, 41–42
Girder bridge:
  preliminary design methodology, 246–249, 252–253
  suggestions for, 104–105
  superstructure
    composite beam and girder bridge, 99–100
    continuous composite-plate girder bridge, 100–103
    plate-girder bridge, 98–99
Golden Gate bridge (U.S.), 26, 27

Hamana-Ko Lane bridge (Japan), 34–35
Harmony, aesthetics and, 169
Height, structural system, geometry, 91
Hell's Gate arch bridge (U.S.), 21, 22
Herodotus, 7
Highway:
  clearance for, preliminary design methodology, 231–232
  construction of, bridge layout and, 53
  crossing for, structural system, geometry, 91
  design methodology assumption, 202–203
  reinforced concrete bridge, preliminary design, theoretical methods of, 219–222
Historical bridges, 1–44
  ancient period, 5–8
  cable-stayed bridges, 30–32
  concrete bridges, 32–35
    prestressed concrete, 34–35
    reinforced concrete, 32–34
  development of types, 2–5
  iron and steel bridges, 17–22
  Middle Ages, 13–16
  Roman period, 9–13
  suspension bridges, 22–29
Howe truss, superstructure (steel bridge), 110
Humber suspension bridge (U.K.), 28–29
Hydraulic conditions, layout conditions, crossing design, 69
Hydrological data, preliminary design methodology, 230

Ice, hydrological data, preliminary design methodology, 230
Ice-breaking cutwater, piers, 152–153
Image system, 301–302
Immortal bridge (Ponte Vecchio bridge, Italy), 14
Impact, specifications and codes, loads on bridge, 183–184
Industrial Revolution, iron and steel bridges, 17–22
Iron bridges, historical bridges, 17–22

K-type truss, superstructure (steel bridge), 110, 111

Layout, 53–61
  angle of crossing, 54–59
  alternatives, 57–58
  generally, 54–56

skewed substructure, 56–57
approaches, 59
crossing design and, 69–73
overview of, 53–54
site investigation, 59–60
field work, 60
general data, 59
office work, 60
Length:
preliminary design methodology, 231
structural system, geometry, 91
Leonardo da Vinci, 4
Lin, T. Y., 39–40, 41
Live loads, specifications and codes, loads on bridge, 180–181
Load distribution, specifications and codes, 187–188
Load intensity reduction, specifications and codes, loads on bridge, 181–182
Loads on bridge, specifications and codes, 179–186. *See also* Specification and codes: loads on bridge
Local conditions:
preliminary design methodology, 228–232
reinforced concrete bridge, alternative comparison example, 265
steel bridge, alternative comparison example, 270, 272
Local scour, crossing design, 76–77. *See also* Scour
Location, 45–52
aesthetic considerations, 46, 167
crossing design
bridge opening design, 66–67
span layout and, 62, 64–65
design methodology, basic parameters, 210–211
technical requirements, 46
traffic flow and, 45, 46–52
generally, 46–50
numerical example, 50–52
transportation requirements, 45–46
London Bridge (U.K.), 13–14
Longitudinal beams, load distribution, specifications and codes, 187–188
Longitudinal forces, specifications and codes, loads on bridge, 184
Longitudinal profile, preliminary design methodology, 228

Mackinac Straits suspension bridge (U.S.), 26, 27
Maillart, Robert, 33, 140
Maracaibo cable-stayed bridge (Venezuela), 30–32, 145

Materials, aesthetics and, 169
Materials cost method, alternative comparison, 261–262
Menai Straits bridge (U.K.), 22–23
Messina Straits suspension bridge (Italy), 36–37
Methodological design trends. *See* Design methodology; Preliminary design methodology
Microcomputer applications, in computer-aided design, 304–305
Middle Ages, historical bridges, 13–16, 202
Minimal wear requirements, design methodology assumptions, 205
Movable bridge, aesthetics, 171

Navigational clearance, preliminary design methodology, 231
Navigation channels, structural system, geometry, 91–92
Niagara Suspension bridge (U.S.), 23–24
Normandy cable-stayed bridge (France), 37–38

Office work, layout, site investigation, 60
Old London bridge, 13–14
Opening. *See* Bridge opening
Ornamentation, aesthetics and, 168
Orthotropic deck bridge, superstructure (steel bridge), 106–109
Outline, aesthetics and, 167–168

Palladio, Andrea, 206
Performance conditions, alternative comparison by, 264
Peter of Colechurch, 13, 14
Pier(s), 151–158
generally, 151–152
girder bridge, preliminary design methodology, 246, 248
with ice-breaking cutwater, 152–153
materials and construction, 153–155
structural system, superstructure and substructure, 86, 87
types of, 155–157
Planning. *See* Design methodology; Preliminary design methodology
Plate-girder bridge:
continuous composite-plate girder bridge, 100–103
superstructure, steel bridge, 98–99
Plongastel bridge (France), 33
Pont du Gard (France), 11
Ponte Vecchio bridge (Italy), 14, 15
Pont Neuf bridge (France), 15, 16

316 INDEX

Portland cement industry, bridge history and, 32–33
Pratt truss, superstructure (steel bridge), 110–111
Precast balanced cantilever, prestressed concrete segmental bridge, superstructure, reinforced concrete bridge, 133
Precast-beam bridge, prestressed concrete bridge, superstructure, reinforced concrete bridge, 148
Preliminary design methodology, 224–260. *See also* Design methodology
  basic work determination during design of alternatives, 253
  general consideration, 226–227
  generally, 224–225
  girder bridge supports, 246–249
  local conditions and construction problem solutions, 228–232
  reinforced concrete, precast simple-span bridge, material expenditure in, 253–256
  reinforced concrete bridge system, 232–237
  sequence of work during, alternatives, 227
  span design, 249–253
  steel and composite bridge, 237–246
  theoretical methods of, 214–223
    final alternative selection (reinforced concrete bridge), 219–222
    generally, 214–218
    practical methods, 218–219
  weight of steel in steel and composite span bridge, 256–259
Prestressed concrete bridge:
  history of, 34–35
  segmental, 132–137
  superstructure (reinforced concrete bridge), 145–149
Proportion, aesthetics and, 167–168, 169–170

Quebec cantilever bridge (Canada), 20–21

Railway:
  bridge history and, 17, 23–24
  design methodology assumption, 202
  future prospects, 40–41
Reinforced concrete:
  abutment material, 163
  seismic design, 194–196
  volume of, steel bridge, alternative comparison example, 278–279
Reinforced concrete bridge:
  aesthetics, 175
  alternative comparison, examples, 265–270
  history of, 32–34

preliminary design methodology, 232–237, 249–250
  precast simple-span, material expenditure in, 253–256
  theoretical methods of, final alternative selection, 219–222
  superstructure, 129–150. *See also* Superstructure (reinforced concrete bridge)
Rialto bridge (Italy), 15, 16
Ribs, orthotropic deck bridge, superstructure (steel bridge), 107–108
Rivers:
  crossing design, 62–85. *See also* Crossing design
  design methodology, basic parameters, 211
  hydrological data, preliminary design methodology, 230
Roebling, John, 23–24, 25
Roebling, Washington, 24
Rolled-beam bridge:
  design suggestions for, 104–105
  superstructure (steel bridge), 97–98
Roman period, historical bridges, 9–13, 202, 203

St. Louis bridge (Eads bridge, U.S.), 18–20
Saint-Nazaire cable-stayed bridge (France), 30, 31
Sando bridge (Sweden), 34
Sant'Angelo bridge (Italy), 11, 12
Scientific research, design methodology assumptions, 208–210
Scour, 76–84
  bridge opening computation and, 81–84
  estimate of, 81
  factors affecting, 77–79
  foundation protection for, 79–80
  generally, 76
  local scour, 76–77
  minimization of effects, 80–81
Segovia aqueduct (Spain), 11, 12
Seismic design, specifications and codes, 190–196
Shaft, pier, 152
Shear, load distribution, specifications and codes, 187
Sidewalk(s):
  clearance for, preliminary design methodology, 231–232
  loading, specifications and codes, loads on bridge, 183
Simplicity, aesthetics and, 168–169
Single-hinged arch, superstructure (reinforced concrete bridge), 138

Site investigation, 59–60
  field work, 60
  general data, 59
  office work, 60
Skew bridge:
  layout, 53–54
  substructure, 56–57
Slab and girder bridge, history of, 33
Slab bridge, superstructure (reinforced concrete bridge), 129
Software (drafting), integration of design and, 308–309
Span:
  design of, preliminary design methodology, 249–253
  determination of, economic principles, structural system, 92–95
  layout of, crossing design, 62–63
  size of, design methodology, basic parameters, 213
  structural system, geometry, 91
Span-by-span construction, prestressed concrete segmental bridge, superstructure, reinforced concrete bridge, 133
Span construction type, design methodology, basic parameters, 213–214
Specifications and codes, 177–197
  earth pressure, 190
  general data sources, 177–179
  load distribution, 187–188
  loads on bridge, 179–186
    dead loads, 179–180
    impact, 183–184
    live loads, 180–181
    load intensity reduction, 181–182
    longitudinal forces, 184
    sidewalk loading, 183
    thermal forces, 186
    uplift, 186
    wind loads, 184–186
  seismic design, 190–196
  substructure, 188–190
Square bridge, layout, 53–54
Starling, pier, 152
Starling coping, pier, 152
Stay cable. *See* Cable-stayed bridge
Steel bridge:
  aesthetics, 170–175
  alternative comparison, examples, 270–295
  continuous composite steel bridge girder, direct optimal design, computer-aided design, 307–308
  historical bridges, 17–22
  preliminary design methodology
    generally, 237–246
    weight of steel in, 256–259
  superstructure, 97–128. *See also* Superstructure (steel bridge)
STEEL DETAILER, 306–307
Steel plate girder, preliminary design methodology, 239–240
Steel weight:
  preliminary design methodology, in steel and composite span bridge, 256–259
  steel bridge, alternative comparison example, 274–278
Steinman, David Bernard, 26
Stephenson, Robert, 17
Stiffening girder or truss, suspension bridge, superstructure (steel bridge), 115
Stonemasonry, abutment material, 163
Straight-wing abutment, described, 160–161
Stringers, load distribution, specifications and codes, 187–188
STRUCTURAL DESIGNER, 305–306
Structural steel, seismic design, 194
Structural system, 86–96
  bridge geometry, 88–92
  bridge types, 95
  span determination, economic principles, 92–95
  superstructure and substructure, 86–88. *See also* Substructure; Superstructure
Substructure:
  abutments, 159–165. *See also* Abutment(s)
  piers, 151–158. *See also* Pier(s)
  skewed, 56–57
  specifications and codes, 188–190
    stream current, floating ice, and drift, 188–190
    wind loads, 185–186
  structural system, 86–88
Superstructure (generally), structural system, 86–88
Superstructure (reinforced concrete bridge), 129–150
  arch bridge, 138–141
  box-girder bridge, 131–132
  cable-stayed bridge, 141–145
  deck-girder bridge, 129–131
  frame bridge, 137
  prestressed concrete bridge, 145–149
  prestressed concrete segmental bridge, 132–137
  slab bridge, 129
  suspension bridge, 140–141
  wind loads, specifications and codes, 184–185

Superstructure (steel bridge), 97–128
  arch bridge, 112–114
  cable-stayed bridge, 118–126
    composite forms, 126
    deck types, 121
    generally, 118
    main girder and truss, 121–124
    stay cable arrangement, 118–119
    stay cable position, 119–120
    structural advantages, 124–125
    suspension bridge compared, 125
    tower types, 120–121
  composite beam and girder bridge, 99–100
  composite box-girder bridge, 105–106
  continuous composite-plate girder bridge, 100–103
    generally, 100–101
    optimum height of plate girder, 101–103
  girder bridge design, suggestions for, 104–105
  orthotropic deck bridge, 106–109
  plate-girder bridge, 98–99
  rolled-beam bridge, 97–98
  suspension bridge, 115–117
  truss bridge, 109–112
  wind loads, specifications and codes, 184–185
Support types, design methodology, basic parameters, 214
Surfacing, orthotropic deck bridge, superstructure (steel bridge), 109
Survey reports, site investigation, field work, 60
Suspension bridge:
  aesthetics, 172–173
  cable-stayed bridge compared, 30, 125
  future prospects, 36, 38–39
  history of, 7–8, 22–29
  structural system, bridge types, 95
  superstructure (reinforced concrete bridge), 140–141
  superstructure (steel bridge), 115–117
Sydney concrete arch bridge (Australia), 34, 35
Symmetry, aesthetics and, 168–169

Technical requirements, bridge location, 46
Thermal forces, specifications and codes, loads on bridge, 186
Three-hinged arch, superstructure (steel bridge), 113
Through-truss bridge:
  preliminary design methodology, 240, 242
  structural system, 88, 89

Tiber bridge (Italy), 33
Timber bridge, design methodology, basic parameters, 212
Tower:
  cable-stayed bridge, 120–121
  suspension bridge, 116
Traffic flow:
  bridge location and, 45
  design methodology assumption, 203
  generally, 46–50
  numerical example, 50–52
Transportation requirements, bridge location, 45–46
Transverse bracing, structural system, superstructure and substructure, 88, 89
Transverse construction, structural system, 86, 87
Truss bridge:
  aesthetics, 171
  superstructure (steel bridge), 109–112
Tubular bridge, bridge history, 18
Two-hinged arch:
  reinforced concrete bridge, 138
  steel bridge, 113

Underpass, clearance, preliminary design methodology, 232
Uplift, specifications and codes, loads on bridge, 186
Urban regions, bridge location, 45–46
U-type abutment, described, 161

Verrazzano-Narrows suspension bridge (U.S.), 26–28, 29
Viaduct, design methodology, basic parameters, 211
Victoria bridge (Canada), 18, 19

Waddell, J. A. L., 53
Walnut Lane Bridge (U.S.), 34
Warren truss, superstructure (steel bridge), 109–111
Water passage data, crossing design principles, 63–65
Width, structural system, geometry, 91
Wind bracings, structural system, superstructure and substructure, 87
Wind loads, specifications and codes, loads on bridge, 184–186
Wing-type abutment, described, 159–160

624.25
T845p

$140.